PUBLIC-SECTOR PROJECT MANAGEMENT

MEETING THE CHALLENGES AND ACHIEVING RESULTS

David W. Wirick, PMP, CMA

John Wiley & Sons, Inc.

Published by John Wiley & Sons, Inc., Hoboken, New Jersey
Published simultaneously in Canada

For general information about our other products and services, please contact our Customer Care Department within the United States at (800) 762-2974, outside the United States at (317) 572-3993 or fax (317) 572-4002.

Project Management Institute (www.pmi.org) is the leading advocate for the project management profession globally. Founded in 1969, PMI has more than 500,000 members and credential holders in 174 countries. PMI's Project Management Professional (PMP®) credential is globally recognized as the gold standard credential in project management

Library of Congress Cataloging-in-Publication Data:

Wirick, David W.
Public-sector project management : meeting the challenges and achieving results / David W. Wirick.
p. cm.
Includes index.
ISBN 978-0-470-48731-0 (cloth)
1. Project management. 2. Public administration. 3. Economic development projects. I. Title.
HD69.P75.W57 2009
352.3'65–dc22

2009018575

10 9 8 7 6 5 4 3 2 1

This book is dedicated to Ellen and Katherine, two people who make a difference in the world, and to those who serve the public interest. There is no higher calling.

Contents

Preface

What do the Great Pyramid of Giza and the Marshall Plan have in common? Both are examples of public-sector projects that had—as do many public-sector projects—an impact on their societies. The Great Pyramid of Giza took 20 years to complete, involved up to 200,000 salaried workers, was constructed to an accuracy of .05 percent, and for 3,800 years was the tallest man-made structure in the world. The Marshall Plan was created in 1947 to rebuild Western Europe after World War II. In its four years of operation, it distributed $13 billion in economic and technical assistance. By the time the Plan was terminated, the economies of every participating nation, except Germany, had grown to prewar levels.

Not every public project impacts society in the same way that these two projects did. But every public project has the capability to improve the lives of citizens or the effectiveness and efficiency of government. Those who manage public-sector projects participate in a proud tradition of projects that created results for society.

Managing projects is always a challenge, especially when those projects involve multiple stakeholders, new or unproven technology, shifting or unclear project requirements, and constrained resources. Those project challenges multiply in the public sector, which depends on successful projects to make the changes necessary to cope with a fast-changing world.

Compounding the challenges of public-sector projects is the shortage of good project managers in the public sector. As the public sector is increasingly challenged to do more with less, it will need the talents of thousands of solid project managers, something public-sector organizations are just now coming to grips with. At the national, regional, state, and local levels, governments are seeking ways to increase the number of project managers at their disposal and build the skills of those they have, so they can manage the complex projects on the public agenda. Other governments are outsourcing the management of projects to private-sector project managers, who will need a deep understanding of public-sector processes and constraints.

That new economy that we hear so much about is the driving force behind the need for better project management. Government across the world is, in fact, facing a perfect storm of increasing demands for services with a decreasing ability to raise revenues. Today's public-sector managers will be required to compete in world markets that will demand much of them and provide them with fewer resources than they need. They will be forced to make their way in an economic environment that is much tougher than that of just a few years ago. The next chapter examines the ways private-sector organizations have tried to respond to this new and challenging environment in an attempt to gather clues as to how public-sector organizations can also thrive.

Project management is one of the best tools for those who work in government or with government and public-sector projects. Although project management is not a magic bullet for success, I would not want to attempt to survive in the new economy or try to deliver results in the public sector without that bullet in my gun. Solid project management skills will not guarantee that public-sector managers will accomplish great and wonderful things, but having those skills will increase the probability that they will.

OBJECTIVES

This book has been designed to address the particular and daunting challenges of managing public-sector projects and increasing project management skills in the public sector. It introduces project management methods and tools that have proven useful in both public-sector and private-sector organizations and identifies both the difficulties of public-sector projects and best practices for dealing with them.

Ultimately, this book is designed to enable public-sector project managers to guide the crucial dialogues necessary for successful projects. It is also intended to help them organize and inspire public organizations to take the actions necessary to create project deliverables that meet organizational needs. It provides those project managers with the hard and soft skills necessary for success in the public sector.

This book follows and attempts to be compliant with the project management methods established by the Project Management Institute (PMI®), the world's largest and most respected association of project managers. It attempts to avoid conflicts with both *A Guide to the Project Management Body of Knowledge (PMBOK® Guide*—Fourth Edition) and the *Government Extension to the PMBOK® Guide Third Edition*, which is also a PMI publication.

This book is for public-sector project managers who want to optimize project outcomes and make a positive contribution to their agencies and organizations. It is for those who have worried about meeting a deadline for a project or getting the support of their stakeholders. It is for those who have wondered about whether they knew what was being requested of them by their managers or stakeholders. It is for those who did not get the results they wanted or who are concerned about new demands and challenges in a high-speed and very demanding environment.

This book is also for those private-sector project managers who are performing projects under contract with the public sector or whose projects interface with public-sector processes or programs. As those private-sector project managers attempt to achieve project results, they may become frustrated with the size of the community of project stakeholders that needs to be satisfied and with public organizations that seem to value rigorous attention to the many constraints that affect public-sector projects over the attainment of business objectives. As will be discussed later, that focus on constraints over results is simply the result of the inherent structure of public-sector programs, which find their roots in statutes, rules, and processes. That focus may also be the result of the fact that penalties for nonconformance outweigh the benefits of attaining results in most public-sector organizations.

No matter what the cause, public-sector projects must be managed differently than most private-sector projects and are subject to additional challenges. The good news, however, is that most public-sector employees are competent and want to achieve outcomes that benefit their organizations and the public. With an understanding of public processes and constraints, private-sector and public-sector project managers can optimize the outcomes of their government projects.

The skills that are embedded in modern project management are survival skills for the future that can be applied in virtually any endeavor. If those skills can be mastered, success will be more likely. In addition, project management skills can be taken from government into the private sector. In fact, one of the principles governing this book is that public projects are tough, and that, if you can manage them, you can manage any projects.

The objectives of this book are to:

- Increase the capability of public-sector managers and private-sector project managers working under government contracts to successfully manage projects
- Create recognition of the importance of good project management in the public sector
- Improve the work products created by public-sector project managers and increase stakeholder satisfaction
- Provide public-sector project management guidance that is consistent with project management best practices, many of which are articulated in the *PMBOK® Guide*—Fourth Edition
- Emphasize the role of planning in order to manage customer, manager, and project team expectations
- Link participants to the traditions of public service and public-sector projects that have had an impact on our world

In order to make participants aware of the long and illustrious history of public-sector project management, a case study of a public-sector project is presented in each chapter. In addition to discussion questions and exercises for each chapter, key public-sector project management terms are included in the glossary.

OUTLINE OF THE BOOK

This preface introduces the book and its goals. Chapter 1 explores the challenges of public-sector project management and the relevance of project management for public-sector managers. It examines the differences between the public sector and the private sector as those differences apply to projects and looks at the looming tsunami about to impact government and how private-sector organizations have coped with the changes that have affected them. It also explores the challenges of project management in the public sector.

Chapter 2 lays the foundations for a study of project management in the public sector. It introduces some of the critical definitions of project management and critical success factors for public-sector projects. It also explores project management maturity models and how they can be adapted to the public sector. Last, it discusses how scalable models of project management can be adapted for a variety of public-sector project types. Chapter 2 begins to explain the need to address public-sector projects as broad communities of stakeholders that require as much management as the details of the project.

Chapter 3 identifies a framework for public-sector project management and details the specific processes, Process Groups, and Knowledge Areas necessary for effective public-sector project management. It identifies the necessary functions for public-sector project management and discusses how public-sector project managers can select the right project management processes for their projects based on the risks of their projects.

The next nine chapters examine the nine project management Knowledge Areas with a focus on identifying the challenges embedded in those knowledge areas for public-sector project management. Chapter 4 begins an examination of those project management knowledge areas with a look at project integration. It examines the functions necessary for integrating the various aspects of the project and creating the project plan. It also examines the management and monitoring of project work in public-sector projects. And it focuses some attention on the initiation of the project, including the creation of the project charter, the principle document necessary for good public-sector project initiation.

Chapter 5 addresses the critical importance of managing project scope and the necessary processes for it in public-sector projects, including the development of the work breakdown structure (WBS). It highlights the WBS as the centerpiece of project planning and the definition of the project's scope. It also addresses the challenges of managing project scope in the public sector.

Chapter 6 describes methods for creating and managing the project schedule. It focuses attention on the creation of the project network diagram and the identification of the critical path. It describes and applies methods for project duration compression and the special challenges of managing the schedule for public-sector projects, in which elements of the project that take considerable amounts of time may be outside the control of the project team.

Chapter 7 describes the role of project cost management, a function that is too often overlooked in public-sector projects. It examines the techniques for estimating and managing the resources and costs of the project and identifies the challenges of public-sector project cost management. It identifies methods for project selection and prioritization and builds a model for prioritizing public-sector projects. It also describes and applies earned-value management as a tool for integrating schedule, cost, and performance data.

Identifying and managing project quality in public-sector projects is the focus of Chapter 8. It examines methods and tools for identifying and managing quality criteria for public-sector projects. It introduces the requirements cycle and the attributes of good requirements as it explores meeting customer requirements as a definition of project quality for public-sector projects. It also describes the importance of compliance with applicable rules and laws as an indicator of public-sector project success.

Chapter 9 is devoted to a discussion of managing project human resources. It examines ways to build and nurture a project team within the constraints of the public sector. It also examines methods for managing public-sector project teams and methods for effectively managing project conflict. Last, it looks at the leadership role of the project manager as he or she attempts to inspire the project team.

Chapter 10 examines the critical role of project communications, arguably the most important role of the project manager. It identifies strategies to determine the communications needs of stakeholders and builds a plan to satisfy those needs. Last, it addresses the emerging need for project managers and other agency managers to adopt strategies to prevent the loss of organizational knowledge as the baby boom generation leaves the workplace and as workforces are downsized.

Chapter 11 addresses the management of project risks. It examines the unique risks of public-sector projects and the tools for project risk management in those projects. It shows how to build a risk register for a public project that contains identified risks, the results of risk analysis, risk-response plans, secondary risks, risk owners, residual risks, and risk triggers. It also looks at the management of the project constraints that affect public-sector projects and identifies a new tool, the constraint management plan, which can be useful for public-sector projects.

Chapter 12 discusses public-sector project procurement and vendor management. It begins with an examination of the project procurement management techniques, which are becoming more critical as public-sector organizations increase their utilization of outsourcing as a means of reducing costs. It examines the specific procurement constraints imposed on public-sector projects. Last, it identifies best practices for managing outcomes and vendors, including the creation of good service-level agreements and statements of work for vendors.

Chapter 13 introduces the role of complexity and chaos in public-sector projects, which tend to be more complex than many of their private-sector counterparts. It identifies and applies a special set of supplementary tools for the management of chaos and complexity in projects. Although good project planning is intended to reduce the complexity of the project environment, in many public-sector projects, outside factors introduce high levels of complexity. This chapter describes and applies three tools for managing those influences.

Embedded in this book is my deeply held appreciation for the work of public-sector employees, who strive in the face of nearly overwhelming constraints to do their best and to serve the public. We have come through a painful period during which public servants were labeled as a problem to be solved rather than as an asset to be fostered. In truth, we have always depended on the efforts of public servants to create and maintain an orderly and just society. As we charge into this new millennium and this new economy, we will need them more than ever before.

The Great Pyramid of Giza

Although a lot of public-sector projects create dramatic results, not every public project creates one of the Seven Wonders of the Ancient World. One that did was the Great Pyramid of Giza.

When the Egyptian pharaoh Khufu decided to create a burial tomb for himself and his close family, he went big. The Great Pyramid we see today is only a small part of the entire structure, which included two temples, three smaller pyramids for Khufu's wives, a raised causeway, and a set of small tombs for nobles. When it was finished, the Great Pyramid became the tallest structure in the world and held that title for nearly 4,000 years until the construction of the Eiffel Tower. What we see today is smaller than the original because the casing stones have fallen off and been moved away from the site. It was composed of 2,500,000 limestone blocks, each weighing between two and 70 tons.

The project team consisted of between 100,000 and 200,000 men, who had to be housed, fed, clothed, paid, coordinated, and managed. Given the length of the project and the hazards of construction, some of those workers had to be buried, and a cemetery for workers has been found. The chief architect (and probably project manager) was

Khufu's vizier, Hemon. There is no record of his earning his PMP. Debate continues as to the construction techniques applied, though they were probably innovative and comparable to the use of new technology today. The project was completed in 20 years, a short time given the size of the Great Pyramid and the daunting tasks that had to be performed.

In addition to the sheer size of the project, it was built to exacting quality standards. Some estimates conclude that construction accuracy was within .05 percent of the plan, which is astounding given the size of the endeavor and the crude tools available. It was also aligned to the four points of the compass, a feat that was regarded as being beyond the capability of the technology available at the time.

Functional requirements were also demanding. The project had to meet the needs of a demanding pharaoh and his family as they made their transition into the afterlife, which required a complex set of internal chambers and passages, including ventilation shafts that maintain the interior temperature of the Pyramid at 68 degrees Fahrenheit. Not only did the structure have to be secure, but it had to be secure *forever* so that the bodies of the pharaoh and his family would never be disturbed. Unfortunately, the project did not adequately assess the ingenuity of thieves, who simply dug around sealed entrances and quickly disturbed the burial chambers.

With an ambitious design, innovative construction techniques, a dedication to meeting demanding quality standards, and the careful coordination of a gigantic project team, the project manager for this public project was able to create a marvel of engineering that has endured for centuries.

Chapter 1

The Challenges of Public-Sector Project Management and the Coming Storm

THE DISTINGUISHING CHARACTERISTICS OF THE PUBLIC SECTOR

Before embarking on a study of public-sector project management, including its unique characteristics, we should first identify how the public sector differs from the private sector. More differences exist between private-sector organizations and public-sector organizations than just their approach to earning and distributing revenue.

Of course, there are differences among public-sector agencies as well. Some public-sector organizations can be defined as public enterprises that are charged with the provision of services on a self-supporting basis. These include municipal utilities that provide water, wastewater, sewer, and other services. Other public-sector organizations can arguably be described as only quasi-public. Examples of these organizations are state-supported universities, which receive an increasingly lower percentage of their operating funds from the states they are in.

Some public-sector organizations provide direct services to the public, although those services are increasingly being outsourced as well. A good example is the provision of mental health services by state institutions. Until the 1970s, state institutions were one of the primary modes of service provision to people with mental illnesses or mental retardation. Since then, those institutions have largely been closed, and service provision has moved to private hospitals.

Other public-sector agencies set standards for the industries or perform economic regulation. Public service commissions at the federal and state levels set rates for gas, electricity, and telecommunications providers. In the past decade, some of those services have been deregulated, and market mechanisms are allowed to set rates. Nonetheless, public service commissions still retain general oversight of the quality of services and the maintenance of effective markets.

Some public-sector organizations are also responsible for ensuring that other agencies comply with the myriad of laws, rules, and process requirements that have been levied on public-sector agencies. Those organizations exercise formal and informal supervision of other agencies and may set requirements for agency operations. Budget agencies not only prepare the budget for the jurisdiction (e.g., the city, the state, the nation) but also are responsible for ensuring that the agencies comply with budget requirements and conform to appropriated limits. These agencies create or enforce many of the constraints that impact public-sector projects.

Despite this array of types of public-sector organizations, they have some shared characteristics, particularly with regard to the management of their projects. Descriptions of those shared characteristics follow.

The Public-Service Purpose

Although they sometimes provide services to distinct populations (like issuing hunting and fishing licenses), all public-sector organizations operate to serve the larger public. That service to the public complicates the management of public agencies and public-sector projects, because it makes identifying objectives much more complex. Not only do a variety of opinions attend the best way to serve that public, but the public itself is difficult to define. For example, what is the goal of a public-sector program designed to revitalize neighborhoods? And who is the public to be served by that program? Is the goal of the program to encourage new investment and development in the neighborhood, which might draw new residents to the neighborhood and consequently drive out current low-income residents? Or is the goal to make housing affordable to current residents? The answers to those tough questions are not without controversy and can substantially impact the direction of the program and the projects within it.

In general, public-sector agencies lack the simple measures of performance, like return on investment (ROI), that private-sector organizations enjoy. Although simple project outcomes, like on-budget performance and timeliness, can be measured, larger outcomes, like the impact on public welfare, are more difficult to measure.

Overlapping Oversight Mechanisms

Public agencies are constrained by overlapping oversight structures. A public agency may operate under (1) the oversight of an elected executive (e.g., a governor or the President), (2) oversight agencies like the Government Accountability Office (GAO) or an office of the budget at the state level, (3) legislative bodies and their own oversight agencies (e.g., a legislative budget office), and (4) elected oversight officials, such as state auditors and treasurers. The constraints of these overlapping oversight agencies are embedded in statutes, rules, executive orders, and required processes. This overlapping oversight represents, at the operational level, the system of checks and balances that limits the power of government agencies to operate outside the bounds of public authorization.

As a result of this overlapping oversight, public-sector projects may be required to dedicate substantial resources to ensuring that constraints are not violated and that oversight agencies are placated. These constraints are, in fact, designed to limit agency discretion and operations so that public-sector employees remain accountable. In addition, the penalties on public-sector agencies for violating these constraints are so severe that public-sector agencies may be very risk averse, even to the extent of choosing compliance over the attainment of business objectives. These overlapping oversight mechanisms also increase the number of project stakeholders with an interest in a project.

For reference throughout this book, it may be useful to identify the hierarchy of official and formal constraints that impact public-sector projects. They are as described in hierarchical order in Table 1.1.

In national governments, the executive is typically called the president; in states, it is called the governor, and in cities, the mayor. In city governments, the legislative body is usually called the city council. For other levels of government, other terms may be applied. For example, in U.S. county governments, the executive and legislative functions are both performed by the county commission made up of elected commissioners.

Table 1.1 Official Constraints of Public-Sector Projects

Formal Constraint	Source	Description	Example
Constitution (or City Charter)	Derived directly from the consent of the people under the authority of their sovereign capacity (i.e., there is no higher authority)	The set of fundamental rules governing a jurisdiction. Constitutions are typically general and describe the powers allowed to the government by the people. Constitutions are deliberately hard to change and are, therefore, entrenched. All specific government actions must be allowed by the general terms of the constitution.	The Constitution of the United States, including amendments made to it since its adoption in 1787
Statute or Law	Created by those elected by the people (legislators or representatives) to act on their behalf	The formal, written enactments of the state. Statutes are created by processes defined by the constitution and under its authority. Laws are enforceable and penalties are specified for violation of laws. Statute law is written and common law is also defined by practice and court decisions.	The Freedom of Information Act (5 U. S.C., 552, as Amended), which makes public records accessible to the public with certain restrictions
Executive Order	Issued by the executive based on either constitutional powers prescribed to the executive or laws passed by the legislature	Regulations issued by the executive under defined constitutional or legislative authority that have the force of law. Executive orders are enforced by courts unless they are deemed to be in violation of the constitution or existing law. Executive orders can be revoked by legislation or superseded by new executive orders.	Executive Order: Facilitation of Presidential Transition, issued October 2009 to describe the roles of persons charged by the President to assist in the transition to the new administration.
Administrative Rule	Issued by public agencies in order to implement statutes	Standards issued by agencies within the powers authorized by statute and	Chapter 123:3-1 of the Ohio Administrative Code, Use of

		which have the force of law. Administrative rules are not subject to the same legislative processes are statutes but must still be issued in conformance with standards that allow for public review and comment. Administrative rules are often issued to create specific guidelines that implement a general statute. Internal agency policies typically do not require the issuance of an administrative rule; requirements that apply to the public or other agencies often do require the issuance of an administrative rule. Administrative rules often require some sort of legislative or executive rule review to ensure that they comply with statute. Administrative rules can be (1) regular rules, which are established through a public comment and hearing process; (2) interim rules, which have a specific time limit; or (3) emergency rules, which can only be adopted in cases of imminent danger to the public. Administrative rules are codified in the administrative code of a jurisdiction.	Electronic Signatures and Records, Office of Information Technology, which prescribes the use of electronic signatures and transactions by state agencies
Mandatory Process or Procedure	Issued by agencies	Describe policies employed internally by agencies to perform their functions. Do not have the force of law and cannot conflict with law or administrative rule.	Agency policy prohibiting the use of e-mail by employees for private purposes

A Short Planning Horizon

Private-sector organizations like to presume that they operate at higher speed than public-sector agencies. Sometimes they do, but there is one area in which the public sector is required to move more quickly. Public-sector agencies have a shorter planning horizon than private-sector organizations because of electoral cycles. Although some public-sector agencies are not subject to election cycles (e.g., publicly supported universities and other board-governed agencies), those that are subject to them are required to articulate an agenda, create plans for implementation of that agenda, and create outcomes in four years, with a four-year grace period if the administration is reelected.

Private-sector, for-profit organizations can establish substantially longer time horizons for product planning and other strategic movements. Public-sector organizations cannot count on the commitment to strategic goals beyond the term of current political officeholders and their appointees.

A Contentious Environment

Every project is subject to conflict and differences of opinion, and private-sector projects may not be supported by all of the organization's stakeholders. But public-sector organizations are subjected to an organized political opposition. That opposition, usually embedded in the opposition party, may be on the alert for opportunities for criticism of the current administration. In addition, the media, though not explicitly attempting to find fault with the current administration, finds "good copy" in the failures of public-sector projects. Unfortunately, failed projects make better stories than successful projects. Both of these factors in combination cause public-sector project managers to feel that they operate in a hostile environment and that they need to avoid visible failure at all costs.

Overlapping Service Delivery Mechanisms

It is rare that any public-sector agency has a monopoly on providing a public service or attaining a public goal. In the United States, for example, services provided to those with mental illnesses may be funded by federal programs and grants, managed by state agencies, and provided by private providers, the state agencies themselves, and county governments. Similarly, education at any level is subject to a variety of funding mechanisms at various levels of government and is provided to the public by an equally extensive array of organizations.

As a result, public-sector agencies have to coordinate their projects with other agencies and consider the impact of their projects on that array of programs and providers. These overlapping service delivery mechanisms also increase the number of stakeholders involved in a project.

Some observers might argue that another difference between public-sector and private-sector projects is that public-sector employees are not adequately motivated. That is not the case. First, though it is true that public-sector employees may not be motivated by short-term financial rewards such as bonuses, they are motivated by the same drives for professionalism and career growth that inspire private-sector employees. Second, they have learned that their motivation for performance must be tempered with an understanding of the constraints under which they work. Blind ambition or revolutionaries cannot be accommodated in public-sector agencies, and public-sector employees have learned that accomplishing objectives requires sharing responsibility and working within existing systems or shaping those systems incrementally.

Third, because of the long-term nature of most public-sector employment and the group cohesion that characterizes many public-sector agencies, public-sector employees have strong group norms and are motivated by a desire to support their colleagues. Although military operations are perhaps an extreme example of public-sector projects, the behavior of soldiers in combat has been shown to be motivated by allegiance to their comrades in small units. Public-sector project managers may want to keep in mind that, in many cases, the allegiance to the small group exceeds the allegiance to the larger agency or organization.

Last, public-sector employees are also motivated by a concern for the public interest. Operationalizing that concern requires complex behaviors given the challenges inherent in identifying the public interest and the actions that must be taken to serve that interest. Inspiring project team members based on their public-interest motivation is, of course, more challenging than awarding them bonuses, which is probably also impossible in public-sector agencies. It is a factor, however, that astute project managers can apply to induce team performance.

THE CHALLENGES OF PUBLIC-SECTOR PROJECT MANAGEMENT

Private-sector project managers like to assume that their work is more demanding than projects in the public sector. They assume that their projects are more complex, subject to tougher management oversight, and mandated to move at faster speeds. Although private-sector projects can be tough, in many cases, it is easier to accomplish results in the private sector than in the public sector.

Public-sector projects can be more difficult than many private-sector projects because they:

- Operate in an environment of often-conflicting goals and outcomes
- Involve many layers of stakeholders with varied interests
- Must placate political interests and operate under media scrutiny
- Are allowed little tolerance for failure
- Operate in organizations that often have a difficult time identifying outcome measures and missions
- Are required to be performed under constraints imposed by administrative rules and often-cumbersome policies and processes that can delay projects and consume project resources
- Require the cooperation and performance of agencies outside of the project team for purchasing, hiring, and other functions
- Must make do with existing staff resources more often than private-sector projects because of civil-service protections and hiring systems
- Are performed in organizations that may not be comfortable or used to directed action and project success
- Are performed in an environment that may include political adversaries

If these challenges were not tough enough, because of their ability to push the burden of paying for projects to future generations, public-sector projects have a reach deep into the future.[1] That introduces the challenges of serving the needs of stakeholders who are not yet "at the table" and whose interests might be difficult to identify. Some also cite the relative lack of project management maturity in public organizations as a challenge of public-sector projects.

---

[1] Project Management Institute, *Government Extension to the PMBOK® Guide Third Edition*, 2006, p. 15.

In addition to these complications, public projects are often more complex than those in the private sector. For some projects, the outcome can be defined at the beginning of the project. Construction projects are one example. For other projects, the desired outcome can only be defined as the project progresses. Examples of those are organizational change projects and complex information technology projects. Although the first type of project can be difficult and require detailed planning and implementation, the second type, those whose outcomes are determined over the course of the project, are regarded as more challenging. They require more interaction with stakeholders and more openness to factors outside of the control of the project team.

Because of the multiple stakeholders involved in public-sector projects, the types of projects the public sector engages in, and the difficulty of identifying measurable outcomes in the public sector, more public-sector projects are likely to be of the latter variety and more difficult. Project complexity and tools for managing complexity and chaos will be discussed later in this book.

As a result of the distinguishing characteristics of public-sector organizations, public-sector projects require the management, not only of the project team, but of an entire community. Little is accomplished in the public sector by lone individuals or even by teams working in isolation. Instead, public-sector projects engage broad groups of stakeholders who not only have a stake in the project but also have a voice and an opportunity to influence outcomes. In public-sector projects, even though the project manager may be ultimately accountable, governance of the project and credit for successes must be shared.

The good news for public-sector project managers is that the community of stakeholders, which may seem to be a burden, can also be an opportunity and a source of resources and support. Many of those stakeholders stand ready to provide help to the project manager as he or she attempts to navigate the constraints affecting the project. Others can be enlisted to support the project, and their authority can make the difference between project success and failure.

THE COMING STORM

In addition to the existing challenges of public-sector projects listed previously, some factors will place soon more stress on public-sector organizations and demand even more emphasis on solid project management. Some of the emerging challenges for public-sector organizations will include:

- Modest or stagnant economic growth
- Globalization and the loss of the industrial revenue base and, increasingly, the service-sector revenue base
- A decline in real wages and pressures for tax reform
- Private-sector practices that pass the corporate safety net back to individuals, who may then look to government for such essential security mechanisms as health coverage
- Difficulty in passing on the need for government revenue to taxpayers and a general loss of confidence in government
- Structural limitations on revenue generation, such as Proposition 13 and property tax indexing
- The redirection of scarce public revenues to homeland security and defense without the imposition of war taxes
- The erosion of public-sector income as entitlement programs drain revenues in response to an aging population
- An age imbalance, with fewer workers in the workforce to support an expanding number of retirees and children
- Longer life expectancy, which further burdens entitlement and health programs
- Increasing costs of health care well beyond the level of inflation
- Long-delayed investments in our national infrastructure, including roads, bridges, and water systems

In combination, these factors constitute a looming storm that will require us to question our assumptions about government operations and services. Doing far more with much less will require new thinking about how government performs its work. It will require more innovation than the development of new services. It will take radical rethinking of what government does and how it goes about getting it done. It will take recognition that the temporary budget reductions required in these tough financial times for government are, in fact, permanent.

Private-sector organizations have already experienced similar stressors, in response to economic concerns and a chaotic environment. Those private-sector organizations are focused on the demands of the competitive market, which requires lean, fast-moving structures and cost reductions. Free flows of capital and the demands for measurable financial performance in the short term, consumer choices, universally available electronic communications, and worldwide labor and capital markets have changed the economic climate for companies. As a result, most private-sector organizations are adopting a short-term planning horizon, embracing the need to shift asset risk to others, and recognizing the need to maintain lean organizational infrastructures.

As private-sector organizations move toward these highly competitive models of operations, they are moving away from traditional operating models. That movement is reflected in the end of the lifetime employment guarantee, reduced employee benefits, and the use of temporary staff and vendors instead of long-term employees. Similarly, highly competitive private-sector organizations are attempting to reduce their reliance on careful processes and procedures. Instead, they are pushing responsibility for decision making to staff at the interface between the organization and the customer.

Newspapers are full of evidence of private-sector organizational transitions. Some of that evidence in the media includes:

- The movement of manufacturing to Central America, Asia, and now Eastern Europe
- Announced layoffs in all industries
- The creation of two-tiered employment strategies (maintenance of pay and benefits for existing employees but lower pay and benefits for new ones)
- The shifting of health-care costs to employees
- Outsourcing administrative functions like IT, accounting, and human resources

These changes have created a set of new organizational strategies, which include:

- Outsourcing
- Cost cutting and downsizing
- The creation of organizations that operate with minimal fixed assets and shifting partnerships with others to exploit network models of organization
- The end of the lifetime employment social contract

- The termination of company-provided benefits and shifting risk back to individuals
- The greater use of temporary and part-time employees

In short, life in the private sector has become less collaborative and more competitive and less controlled and more chaotic. Ask almost any worker in a modern U.S. private-sector organization, and they can tell the same story in vivid detail.

As noted, many of the same pressures that have driven private-sector organizations to adopt the listed strategies will soon impact government and the public sector. Agencies will need to compete for ever-decreasing amounts of revenue, governments will try to create lean government as a means of competing with other jurisdictions for jobs, and demands on government will increase as the social safety net erodes. In short, public-sector organizations will need to adopt some of the same strategies that private-sector organizations have already made as those public-sector organizations face increasing resource constraints and new demands for services. Those changes will be difficult for the public sector.

For decades, public-sector organizations have emphasized organizational models that value stable processes and an aversion to risk. In addition, public-sector compensation systems have valued longevity, and retirement systems have provided great benefits in the future in return for less compensation in the short run. As a result, public-sector organizations have not been structured to be flexible and innovative, two requirements of organizations in the new economy. Whether government agencies want to make the transitions demanded by the coming economic storm, environmental conditions are certain to push them there.

NEW TOOLS FOR PUBLIC-SECTOR MANAGERS IN THE NEW ECONOMY

As public-sector agencies make the necessary transitions to cope with the demands of the new economy and the impact of the factors described earlier, public-sector managers at all levels of government will face an array of daunting challenges. Some of those are:

- Motivating employees who are coping with increasing demands but less pay and security

- Dealing with a multigenerational workplace (According to some observers, there is a wider age range among employees in the workplace now than at any time in history.)
- Managing for short-term results with limited resources
- Managing employees who are not in the same geographic location
- Managing vendors who may be performing critical organizational functions
- Building organizational loyalty without the trade-off of a guarantee of long-term employment
- Managing in an environment of constant change
- Coping with the unique constraints of public-sector organizations, which include political systems, organizational stovepipes, and limited technology
- Coping with the loss of organizational knowledge as the baby boom generation retires

Public-sector managers will require new tools and strategies for operating in this challenging new environment. So what tools can public managers apply? One of the best adaptive tools for organizations and individuals is project management, which is the focus of this book and which is ideal for organizations attempting to create change and optimize the use of scarce resources. In order to make their projects successful, public-sector project managers will need a combination of humility and patience coupled with dogged persistence and creativity. Management tools and skills for making public-sector managers successful will be discussed later in the book.

DISCUSSION QUESTIONS

1. What are the implications of the changes in the workplace? Is the workplace more stressful? Are resources less available? Are you being asked to do more with less? Is it harder to separate work and home? When will things change?
2. How can you build a sustainable work life—one that you can sustain for the long term?
3. What management tools do you think will be effective in the public sector in the future?

EXERCISE

1. Identify a public-sector organization. Identify the pressures (e.g., financial, competitive, technological, workforce) that it might be facing. Create a list. For the pressures on that list, create a second column identifying strategies (projects) for coping with those challenges.

Project Apollo

Some projects discover midway through that they have ambitious and nearly unachievable goals. Some projects start out that way. Some historians today regard Project Apollo and the moon landing as the greatest feat in human history. Project Apollo was second only to the construction of the Panama Canal as the largest, nonmilitary technological project ever performed by the United States.

Amid concern that a "missile gap" had been opened by the Soviet Union over the United States, the United States was particularly concerned when the Soviets put a man in space. On May 25, 1961, President John F. Kennedy responded by declaring that " . . . this nation should commit itself to achieving the goal, before this decade is out, of landing a man on the Moon and returning him safely to the Earth." That goal was particularly bold in that, at the time of Kennedy's speech, the United States had put only one person into space (for less than 16 minutes) and none into orbit around the earth.

Project Apollo, as the project to put a man on the moon was known, was greeted by some as impossible. The head of Kennedy's space flight advisory committee believed that launch-vehicle technology was poorly developed and that putting a man in space was a high-risk endeavor with little chance of success. Ultimately, Project Apollo required 400,000 project team members, and a partnership between the government, universities, and private companies. Kennedy also publicly asked the Soviets to work with the United States in developing space technology, which asserted the United States was at least equal to the Soviet Union in space technology.

Project Apollo accomplished President Kennedy's goal on July 20, 1969, when Neil Armstrong stepped off the lunar module onto the surface of the moon. In order to accomplish its goal, the project identified a set of product-based and project-based deliverables that included the mission structure, the spacecraft, the lunar module, the boosters, unmanned and manned missions, the moon mission itself, and post-mission applications. The project not only accomplished its goals on time, but it performed well from a budget perspective. With an original estimate of about $20 billion, the total cost of the Apollo project was estimated to have been between $20 billion and $25 billion. At the height of the project, the budget of the National Aeronautics and Space Administration (NASA) was 5.3 percent of the total federal budget. A major portion of NASA's project budget (80 to 90 percent) was directed to contractors who created the goods and services necessary for the project.

The project had its challenges and setbacks. When President Kennedy was assassinated in 1963, President Lyndon Johnson took up Kennedy's challenge to the nation and successfully guided NASA appropriations through Congress. Three astronauts died in a launchpad fire. As made famous by a movie of the same name, Apollo 13 experienced an in-space explosion that nearly caused the death of its three-man crew. Engineering setbacks caused the scope of the mission to be adjusted on several occasions, and a shrinking budget caused the last three lunar-landing missions to be cancelled. Project managers had to coordinate the work of engineers and scientists, who differed in their approach and outlook, and the work of contractors, university researchers, and employees.

The project risk was heightened by the significant political and public interest in the project. Not only was the prestige of the United States at risk, but when Apollo 8 orbited the moon—the first time an earth-launched spacecraft had orbited another celestial body—the television audience was the largest up until that time. When Apollo 11 landed on the moon and Neil Armstrong made the first step onto the moon's surface, somewhere between 600 million and 1 billion people watched.

As significant as the results of the Apollo project were, it also had an impact on our understanding of project management and our capability to manage huge endeavors under aggressive time limits.

Chapter 2

The Foundations of Public-Sector Project Management

Increasingly, succeeding in any endeavor is a matter of managing projects. The truth is that private-sector and public-sector organizations make progress and create changes through projects, which are temporary endeavors designed to produce unique products, services, or results. There are few more sought-after skills in the public-sector workplace than project management.

Any effort that is creative is a project. Any attempt to change something or create a new outcome is a project. Launching a military campaign or a run for public office is a project. Creating a painting or a sculpture is a project. Making a movie or writing a book is a project. Merging two agencies is a project. In government, a budget is a project, process improvement is a project, and creation and deployment of a new program are projects. All a project takes is a start, an end, and a unique outcome, and even though projects have different dimensions and challenges, as discussed later, they can be managed using the same principles.

THE PROBLEM WITH PROJECTS

If you think that you and your colleagues have trouble managing public-sector projects, you are not alone. The fact is that many (some even say most) projects fail in both the private and public sectors. Public-sector projects come in over budget, sometimes by massive margins. (The Big Dig in Boston was planned to cost about $3 billion. The actual cost was over $14 billion, and the legal costs are still mounting.) Projects get finished late, and—more often than you might believe—they do not get finished at all. A project designed to produce a new information system for the FBI was abandoned after years of investment and work. (More information about that project is included in Chapter 6.) That is not so odd; many projects get cancelled in the middle or, worse, just disappear without a trace.

Even more often, projects do not deliver what they were intended to deliver. The space shuttle, which had been the result of a massive project, once crashed on liftoff and once crashed on reentry. Floors in new buildings crack and roofs leak. Information system projects are legendary for not meeting the needs of their stakeholders. In fact, the vast majority of failed information technology projects fall short because of differences of opinion between users and developers about what the project was supposed to create.

WHY DO PUBLIC-SECTOR PROJECTS FAIL?

Project-sector projects fail for all of the normal reasons that any project fails. Projects in all sectors of the economy fail because they:

- Fail to identify the needs of customers or users of the product or the project
- Create overly optimistic schedules and fail to anticipate the impact of late deliverables
- Do not get the resources necessary to complete the project
- Do not devote enough time to project planning
- Are subject to changing management priorities
- Employ technology that does not work as expected
- Do not get good performance from vendors
- Get overwhelmed by competing projects and do not apply solid project prioritization
- Do not adequately identify, analyze, and address project risks
- Make assumptions that are not validated and agreed to
- Dissolve in the face of conflict among stakeholders
- Get overtaken by unexpected events (More will be said in Chapter 14 about the challenges of managing uncertainty and chaos.)
- Do not apply solid and repeatable project management methods
- Do not have the benefit of an experienced project manager
- Do not engage and involve stakeholders throughout the project
- Do not identify lessons learned from prior projects
- Define an overly broad project scope that cannot be well-defined

In addition, public-sector projects can fail for a set of reasons related to the unique character of public-sector projects. In that regard, they:

- Run afoul of political processes
- Lack the necessary resources because of requirements to use existing staff rather than to contract for the right expertise
- Are constrained by civil-service rules that limit assignment of activities to project staff
- Lose budget authorization
- Lose support at the change of administration due to electoral cycles
- Are overwhelmed by administrative rules and required processes for purchasing and hiring
- Fail to satisfy oversight agencies
- Adopt overly conservative approaches due to the contentious nature of the project environment
- Are victimized by suboptimal vendors who have been selected by purchasing processes that are overly focused on costs or that can be influenced by factors that are not relevant to performance
- Are compromised by the bias of public-sector managers and staff toward compliance over performance
- Fail to identify project goals given the wide array of project stakeholders in the public sector and the challenges of identifying public-sector goals and metrics for success

If this all sounds daunting, it should. Doing projects right in the public sector requires more than knowledge of project management methods. It requires creativity, communications, organization, conflict management, and hard work. It requires management of the unique constraints imposed by public-sector organizations, which will be discussed at length later As projects get more complex and as the number of people impacted by the projects grows, those projects will become even more challenging. Fortunately, there is hope.

THE GOOD NEWS ABOUT PROJECTS AND PROJECT MANAGEMENT STANDARDS

The good news about project management is that it can be learned. That does not mean that anybody can become a *great* project manager. Being a great project manager requires knowledge of project management, knowledge of the technology being employed, and the ability to communicate, empathize, and manage conflict. Anyone, though, can become a *better* project manager.

Because all sorts of organizations recognized that they could not keep shooting themselves in the foot by failing at project management, they began to try to get better at it. They started studying project management, buying project management training for their employees, and developing project management methods and tools. Good project management became a competitive advantage for private-sector organizations, and project management began to be viewed as a necessary skill for managers at many levels of the organization. Project management also became a tool for reducing the risk of failed projects and, as a result, for contributing to the success of the organization.

Soon after, project management became a recognized discipline in organizations, universities, and professional development programs. The list of universities offering degrees in project management or project management concentrations within other programs, like MBA programs, is growing. And, pulling it all together, a professional association of project managers seized the bull by the horns and developed a worldwide standard in project management and a certification for those who have demonstrated competence in project management. The project management method developed by the Project Management Institute (PMI®) is that standard, and its Project Management Professional (PMP®) Certification is the gold standard in project management certification. The PMI project management Knowledge Areas are described in *A Guide to the Project Management Body of Knowledge (PMBOK® Guide*—Fourth Edition), which is available from PMI and other bookstores.

Recognition of the importance of project management in the public sector has also grown. PMI has also recognized the value of government project management and has done two important things:

1. Established a government project special-interest group to allow public project managers to share best practices and develop their skills
2. Created a *Government Extension to the PMBOK® Guide Third Edition* which provides information on project management within the unique environment of the public sector and accumulates good practices that have become widely accepted in the field

Public-sector organizations have also recognized the value of good project management, as indicated by:

- The creation of project management training programs at the state and federal levels
- The increasing recognition of the PMP® certification as a necessary qualification for public project managers and minimum requirement for some positions

As a result, project management has become a legitimate profession and career path. As a profession, it has created:

- Recognition by employers of the importance of project management and demand for skilled project managers
- Academic courses and recognition
- Certification
- Research (see the *Project Management Journal*, also a publication of PMI)

The remainder of this book will explore project management methods that you can apply in your projects. They are scalable, which implies that they can be applied to simple public-sector projects or to large and complex projects.

THE VALUE OF PROJECT MANAGEMENT TO A PUBLIC-SECTOR ORGANIZATION

Project management can help a public organization adapt to the changing dynamics of the environment within which those organizations must function. As indicated earlier, it is probably the ideal management tool for coping with the storm about to come. Project management works under those circumstances because it:

- Defines outcomes to be achieved
- Identifies the minimal resources necessary to complete those objectives
- Allows the organization to move quickly and to hit schedule targets
- Manages the risks of goal achievement

- Focuses the organization's attention on the goal and the required outcomes
- Creates change

In the public sector, creating a focus on outcomes is particularly important because of the challenges public-sector organizations face in identifying and reaching definable goals. Project management forces dialogue to identify what the organization is trying to achieve and how it intends to achieve it.

One of the most useful attributes of disciplined project management is its ability to make managers and team members engage in "adult thinking." Adult thinking is different from "magical thinking," which many organizations and managers like to engage in. Magical thinking allows a manager or an organization to imagine that they do not have to make choices; magical thinking implies that an organization can have things two ways at the same time. For example, organizations or individuals engaged in magical thinking could act as if:

- The decision to add project scope or remove resources from the project will not have any impact
- Projects can simply be added to the project inventory without having an impact on the projects already pending
- People can work unlimited numbers of hours to accomplish new priorities
- Projects do not need to be prioritized because "it all needs to be done"
- Personnel who do not have the necessary skills will still get the job done
- Projects can be completed despite the unrealistic constraints placed on them
- Project managers will define the project perfectly without guidance (i.e., managers can remove themselves from accountability for projects by failing to define them)

Public-sector managers, faced with many competing demands and stakeholders, are especially prone to levying new demands on staff without eliminating existing demands or prioritizing requirements.

Project management requires defining priorities and needs. It requires identifying needed resources and dedicating them to the project to achieve the desired results. It does not mean that managers cannot change priorities or projects; it simply means that the implications of those changes must be clearly understood. At the same time that project management assigns accountability for performance, it protects the project manager and project team by forcing a clear description of the project and a dialogue about expectations.

THE DOWNSIDE OF PROJECT MANAGEMENT

Despite its being a potent tool to improve the operations of public-sector organizations, project management is not perfect. Its downsides include the facts that:

- Project management can encourage short-term thinking in that the focus on task accomplishment that project management provides is a strength and a weakness.
- Project management can require new skills that may be difficult for some managers to adopt.
- Project management can increase stress by requiring accountability and performance, although it can reduce stress by ending the magical thinking that public-sector managers might have engaged in.
- Project management can provide an incentive for focusing on measurable, tangible benefits ahead of long-term strategic needs.
- Project management, poorly applied, can create new management processes that do not add enough value to offset their cost.
- The deployment of project management processes and methods may irritate senior project managers who prefer to "do things their way."
- Project deadlines can wear out staff.

Too much project management discipline and too many processes can be applied to the extent that the benefits of project management are no longer worth the costs. To avoid that overcommitment to project management, project management processes must be scalable based on the risk of the project. That is to say that if a project is of low risk (e.g., it is a type of project that the organization has had substantial success with or can expect easy success with), the project management methods applied to it do not need to be extensive. A little project management discipline may add value, but extensive processes will not. (Methods for creating scalable project management based on the risks of projects are discussed later.)

For high-risk projects, the project management methods of the organization need to be applied with greater discipline. The purpose, after all, of creating any process is to reduce risk. The same is true of project management methods and processes; they are designed to reduce the risk of project failure. If that risk is high, extensive processes and methods should be applied. If that risk is low, more informal project management methods can be employed.

THE CRITICAL SUCCESS FACTORS FOR PUBLIC-SECTOR PROJECTS

Best practices can increase the probability of project success. Some general project critical success factors for public-sector projects include:

- A project management methodology built specifically for the needs of the projects of the organization that is scalable based on project risk
- An interactive dialogue among stakeholders that continues throughout the process
- A detailed process for identifying user and supplementary requirements
- Management support of the project management process
- Capable project managers with both hard and soft skills who have the ability to:
  - Envision the project as a community endeavor and enlist a broad group in problem resolution
  - Share credit for success
  - Manage complex processes that may be required by law or administrative rule
  - Respond quickly and positively to adversity, which is a constant of public-sector projects
  - Manage conflict among stakeholders and to recognize the interests of even those who might oppose the project
  - Deal with the press when necessary

In later chapters, best practices for management of each of the project management knowledge areas will be identified.

PROJECT MANAGEMENT MATURITY MODELS IN THE PUBLIC SECTOR

A variety of project management maturity models can be used to assess the capability of the organization to manage projects and the extent to which the organization deploys solid project management methods. Those models rate the organization on a 1-to-5 scale, where a "1" indicates ad hoc, poorly developed project management process deployment, and a "5" indicates an organization that applies very sophisticated methods and techniques. NASA is a public-sector organization rated at a level five.

NASA has invested heavily in its project management methods because:

- All of the major work of NASA is accomplished within projects (e.g., Project Mercury)
- NASA projects are highly visible, subject to government oversight, and dangerous
- NASA projects have clearly defined outcomes that can be optimized by good project management
- NASA projects are very complex

Not all public organizations should aspire to high levels of accomplishment on maturity models. Improving project management maturity, particularly at the highest levels, can be expensive and may not be worth the investment. Most public-sector organizations can benefit from improvement, however. For public-sector organizations, it may be appropriate to complement the other factors measured by the maturity models with such characteristics as:

- The extent to which project management is embraced as a strategy by multiple levels or branches of government
- The extent to which government purchasing processes and hiring processes recognize the value of project management and project management certification
- The extent to which regulatory and oversight processes mandate the application of professional project management
- The application of project management to initiatives not typically thought of as projects (e.g., budgeting)

SCALING PROJECT MANAGEMENT METHODS

Not all projects require detailed project management methods, and unnecessary processes that are a burden to project managers should be avoided. The goal is to create project management processes and methods that provide cost-effective reductions of project risk. That is, the goal is to find a set of project management strategies that increase the probability that projects will be successful. That is usually defined as being on-time and on-schedule, with the necessary functionality the product of the project requires. How to achieve each of these outcomes will be discussed later.

The best way to make certain that project management methods are cost effective is to identify the risks of the project types engaged in and create a set of project management methods that should be used for each risk type.

That is a two-step process:

1. Identifying the project management methods to be applied to the various risk classes
2. Analyzing the risks of projects and fitting them to the risk classes

Table 2.1 illustrates how an organization might categorize its projects and project management methods.

Table 2.1 Project Management Methods Applied to Risk Categories

Project Risk Category	Project Management Methods
Category A	Short-form project charter and monthly short-form status report
Category B	Project charter, WBS, risk register, and monthly short-form status report
Category C	Project charter, formal requirements elicitation process, WBS, network diagram, risk analysis and risk register, monthly status report, communications plan
Category D	Project charter, formal requirements elicitation process, WBS, network diagram, budget, earned value analysis, risk analysis and risk register, qualitative and quantitative risk analysis, communications plan, procurement plan, detailed monthly status report, and stakeholder meeting

Categories can be added or deleted and methods tailored to fit the needs of the organization. In any event, the principle remains the same—project management methods should be scaled to the needs of the organization.

Table 2.2 provides an illustration of a project risk analysis tool that can be used to place each project into one of the aforementioned categories listed.

Last, the scores received from the analysis of project risk will need to be correlated with the risk categories. For example, a project scoring between

Table 2.2 Project Risk Analysis

Risk Category	Description	Project Scoring
Number of stakeholders	Mostly internal, few stakeholders	1-3 points
	Some external stakeholders	3-5
	Many external stakeholders	5-10
Project size	Less than 2 months in duration and less than 10 FTEs engaged	1-3
	2-6 months in duration and 10-30 FTEs engaged	3-5
	More than 6 months in duration and more than 30 FTEs engaged	5-10
Public interest and impact	Little impact on the public and little interest by the press or public officials	1-3
	Some public interest and impact, some press and public official interest	3-5
	Significant impact on the public and significant interest by the press and public officials	5-10
Use of technology or methods	Well-known technology that has been applied in the past	1-3
	Some new technology or methods, organizational unfamiliarity with the solution	3-5
	New technology, organization has to rely on vendors to interface with or apply the technology	5-10
Strategic impact	Little impact on strategic goals or outcomes	1-3
	Project is integrated into strategic directions and goals	3-5
	The project is directly involved in the organization's strategic goals	5-10

40 and 50 points might be regarded as a Category D project, whereas a project scoring between 30 and 40 points might be regarded as a Category C project.

THE USE OF SOFTWARE FOR PROJECT MANAGEMENT

This book does not address the use of project management software, though a variety of software exists that can help a project manager track details of the project. For large projects, the use of software is almost required. A few comments about project management software may be appropriate.

First, no software manages a project. People manage projects, and the use of software can create tunnel vision for project managers. Instead of being actively engaged with project stakeholders, project managers can become focused on the software and conclude that the software models are the project.

Second, software has the capability to assist the project manager in two ways. The first, which is the purpose of most project management software, is to help the project manager deal with the high volume of data. Project management software can create a schedule, track performance, and create project status reports that might have been beyond the capability of the project team. The second use of software in project management might be even more important. A new generation of software allows the project team to communicate easily, share documents, and engage in disciplined brainstorming. This software (e.g., Microsoft's SharePoint) allows web portals to be established for a project team and its stakeholders. Access can be controlled to that project site, and stakeholders can use the site to share information and store current versions of key documents.

Third, enterprise project management software can be purchased and applied to coordinate the management of projects across a large enterprise. This software can allow senior managers to follow project status across the enterprise from a management "dashboard." Enterprise project management software can also be used to coordinate and control resource utilization across the enterprise. For best results, consult with an expert in enterprise software deployment prior to attempting to install one of these systems to ensure that your organization is ready to employ the many features of these systems.

DISCUSSION QUESTIONS

1. What causes your public-sector projects to fail? How would you know if a project has failed?
2. How could project management improve your agency or the agencies you are involved with? Is it worth the investment?
3. What examples of magical thinking have you seen? How can that magical thinking be prevented?
4. What are the downsides of project management in your experience? What kinds of reactions would you have to the introduction of project management methods into your organization?
5. In addition to those listed in the chapter, what other critical success factors can you imagine for public-sector projects?

EXERCISES

1. For your agency or a public organization you are familiar with, create a risk analysis framework like the one described in the chapter. Identify the project management methods you will apply to each project risk class.
2. Create a five-step process for measuring the project management maturity of the agencies that you work with. Rate your agency and the agencies they interact with.

The Creation of the Peace Corps

One of the distinguishing factors in public-sector projects is the required involvement of a wide array of stakeholders. With so many stakeholders who can influence the project's outcomes, project managers need to engage those stakeholders and allow them to be involved in the co-creation of the project. Nothing illustrates that more than a project designed to create a new public program.

Although the idea for some form of public service for the nation's young people predated President Kennedy, its direct origins arose

(*continued*)

from extemporaneous comments made by presidential candidate Kennedy to a crowd of students at the University of Michigan in October 1960. Addressing 10,000 students early in the morning, the President challenged them to devote some portion of their lives to public service in Asia, Africa, and Latin America. He asked them if they would support his establishment of a peace corps. Their response was immediate, and upon his election but before his inauguration, Kennedy directed Sergeant Shriver to prepare a feasibility study of his idea.

The forerunners of the Peace Corps included the missions sent by many religious organizations to build schools and teach trades, and, prior to 1960, bills were introduced into Congress to create a volunteer service organization. Those bills did not pass but caught Kennedy's attention, and he established the Peace Corps by Executive Order in 1961 and funded it using State Department funds. Shriver was appointed as Director. At every opportunity, Kennedy and Shriver pushed the program and made it visible.

Eventually, the program would need Congressional support, and requesting that support would require that it face its critics head-on. Kennedy and Shriver were able to make changes in the program that silenced critics. In response to those who were suspicious that the Peace Corps would become a haven for draft dodgers, they agreed that Peace Corps service would grant a deferment but not a draft exemption. To silence critics who were concerned that the Peace Corps would become a tool of the CIA, they agreed that volunteers would only be sent to countries that requested them. In addition, previous work for any intelligence agency was established as an automatic disqualifier for Peace Corps volunteers.

The Peace Corps began as a personal challenge by Kennedy. It was seeded by the overwhelming youth response to that challenge, and it was given energy by a Director and President who were able to find a creative way to start the program. Eventually, the program was made permanent by the willingness of its founders to adapt it to meet the demands of its critics. Kennedy had a special bond with Peace Corps volunteers, who came to be known as "Kennedy's kids" and who were invited to the White House often.

Chapter 3

The Framework for Managing Public-Sector Projects

THE PROJECT MANAGEMENT FRAMEWORK FOR PUBLIC PROJECTS

We have already defined a project: a set of activities with a start and a finish designed to create a unique product or service. Public-sector projects come in all shapes and sizes and can include such endeavors as a(n):

- Military campaign
- Negotiation of a peace treaty
- Lawsuit
- Budget
- Program creation or deployment
- Adoption of a statute or rule
- Process improvement
- Program evaluation
- System deployment or upgrade
- Construction of a building or monument
- Environmental impact study or reclamation
- Agency reorganization
- Issuance of a complex permit
- Financial audit of another public-sector program
- Attempt to attract business to a political jurisdiction

Fortunately, the project management frameworks established by the project management profession and those articulated here are applicable across all project types, including public-sector projects. Some tailoring is required, but the fundamentals work across all project types and disciplines.

GROUPING PROJECTS FOR BETTER MANAGEMENT

In order to improve our ability to manage public-sector projects, we may want to bundle projects. We do that by creating programs and portfolios. Programs are groups of related projects. For example, a program consisting of multiple projects might be created to change the way an agency operates. Projects in that program might include upgrading systems, improving processes, hiring new staff, training existing staff, and reorganizing the agency. Typically, a program manager would be responsible for oversight of the program.

Portfolios are groups of unrelated projects. For example, an agency at any point in time might have initiated a wide array of projects. They might, for example, have information technology projects, organizational change projects, process improvement projects, projects that implement statutory requirements, and projects that create new programs. Although those projects might be unrelated in their outcomes, they probably rely on the same resources and might need to be prioritized and coordinated.

Some public and private organizations use a project management office or organization (PMO) for project, program, and portfolio management. PMOs are usually designed to provide coordinated management of the various projects in the organization, but they can differ widely in their missions. Some are responsible for management of all the projects of the organization, whereas some merely assemble project management resources, and others are engaged in project prioritization. Some PMOs see their role as the deployment of enterprise-wide project management software and methods.

Care needs to be taken in the establishment of PMOs. They are not panaceas for improving project management. If the mission of the PMO is not clearly defined, if it is not given the support of management, if the organization is not ready for centralized control or coordination of projects, or if the PMO does not prove itself and quickly generate demonstrable results, the next budget cutback will probably mean the end of the PMO.

In the public sector, PMOs are often established in oversight agencies, such as central information technology offices. Those PMOs run the extra burden of being regarded by operating agencies as unnecessary overhead. Often, operating agencies regard any central control as a threat to their autonomy and attempt to circumvent the mandates created centrally. PMOs can suffer the same fate and the same loss of credibility simply because of their association with central administration.

Public-sector PMOs are most successful when they:

- Have clearly defined goals
- Operate outside of the information technology office
- Are supported by senior managers
- Delay the deployment of enterprise project management software until the organization develops some project management maturity
- Are staffed with proven experts who know the business functions of the agency
- Demonstrate their ability early with quick wins and visible successes

Any time public-sector agencies attempt to group projects or centralize project management, they are in for a challenging ride. Establishing project priorities is a challenge in any organization, which is why many public-sector organizations have avoided it. The coordination of project resources across a public-sector organization is a challenge as well. The best program and portfolio management systems require resource sharing, which is a challenge in any organization and a particular challenge in hierarchical organizations that typify the public sector.

BREAKING PROJECTS INTO COMPONENTS

In addition to combining projects for better coordination, on most public-sector projects, better management results can be achieved by breaking the project into smaller pieces. Think about, for example, how the Apollo Project described earlier was subdivided to make it more manageable.

Projects can be broken down into phases. Project phases are collections of project activities that usually create a deliverable. A phase should end with the creation of a deliverable so it is clear that the phase has been completed and its outcomes can be evaluated. For example, in a project designed to build and deploy a new government accounting system, the following phases could be created:

- Budget and approval phase
- Needs identification phase
- Design phase

- Development phase
- Test phase
- Deployment phase
- Assessment phase

At the end of each of these phases, the project team and stakeholders can evaluate the deliverables created and decide if changes are necessary. It is even possible that the project might be terminated at the end of a phase if there is not a compelling reason to continue.

For similar projects and those that are likely to be repeated, a project life cycle can be created. A life cycle is simply a set of standard phases that can be used for similar projects. For example, a government department responsible for acquiring and retaining jobs and interacting with private industry might create a life cycle for a business development project. That life cycle would form the basis of project planning and would be used (and modified if necessary) for all business development projects. Phases in an economic development project designed to attract business to a region might include:

- Initial investigation of the opportunity
- Detailed analysis of company needs
- Assembly of a package of incentives
- Presentation of the package to the company and negotiation
- Approval of the final package and contract
- Sign-off and awarding of the incentive package
- Project closure and follow-up

That department could also establish one life cycle for large-scale efforts and a simplified one for smaller business development projects. The objectives are to be able to better manage projects, establish a common vocabulary, and create planning artifacts that are reusable across projects.

The next logical way to break projects into pieces to make them more manageable is to identify the deliverables of the project. Deliverables are the nouns of projects—the things the project will create. Some of those things are created only for the use of the project, like the project plan. Other deliverables become parts of the deployed solution, like training manuals or deployed software.

Deliverables are the base for good project planning. Without a clear articulation of the deliverables of the project, schedule or cost requirements cannot be identified. If a project is begun by identifying the activities for the project, without first identifying the deliverables that those activities will contribute to, it will create a recipe for project failure. The goal of project management is to make sure that every activity is associated with a deliverable. If an activity cannot be clearly associated with a deliverable, it is outside the scope of the project.

Activities are the verbs engaged in to create the deliverables that have been identified. Sometimes, those activities are called tasks. Deliverables can be broken down into layers and activities assigned only to the bottom layer of deliverables, which are called work packages (this will be discussed more later).

Table 3.1 lists a set of deliverables and the activities that would be undertaken to create them for an agency budget request, a common public-sector project. The actual listing of deliverables and activities would be significantly more extensive than the illustrative examples here. Note the difference between the nouns and verbs.

Table 3.1 Illustrative Deliverables and Activities for an Agency Budget Request

Deliverables	Related Activities
Budget schedule and resources	-Identify budget filing deadlines -Identify resources necessary -Determine resource availability -Create a schedule -Circulate the schedule and resources for internal review by managers -Document the schedule -Update the schedule with changes as necessary
Identification of agency needs	-Survey managers to identify program needs -Evaluate expenditure patterns -Document performance information -Prioritize agency needs
Presentation of identified needs to stakeholders and advocates	-Identify stakeholder groups -Identify actions desired for stakeholders -Prepare presentations

(*continued*)

Table 3.1 *(continued)*

Deliverables	Related Activities
	-Deliver presentations -Evaluate presentations and make budget changes(if necessary)
Budget preparation and submission	-Prepare budget forms -Check forms for errors and omissions -Submit budget
Budget negotiation and modification	-Meet with oversight agencies -Present arguments for budget -Present options for changes -Make changes
Budget implementation	-Enter budgets into accounting systems -Notify managers of approved budgets and limits -Monitor performance against the budget -Provide feedback to managers -Recommend changes and corrective actions -Create historical information for next budget cycle

PROJECT PROCESS GROUPS

In order to further organize the work of the project, for every project or every project phase, we can apply what was commonly known as the IPECC model, which consists of project Initiation, Planning, Execution, Control, and Closing. (In some descriptions of that model, "control" is referred to as "monitor and control.")

Those five *process groups* flow as shown in Figure 3.1.

Although these process groups appear to be linear, movement back and forth through them is common. For example, it is very common to find during execution that the project plan has to be changed to deal with emerging conditions. Defects identified during monitoring and control can require a return to project execution in order to correct errors in deliverables.

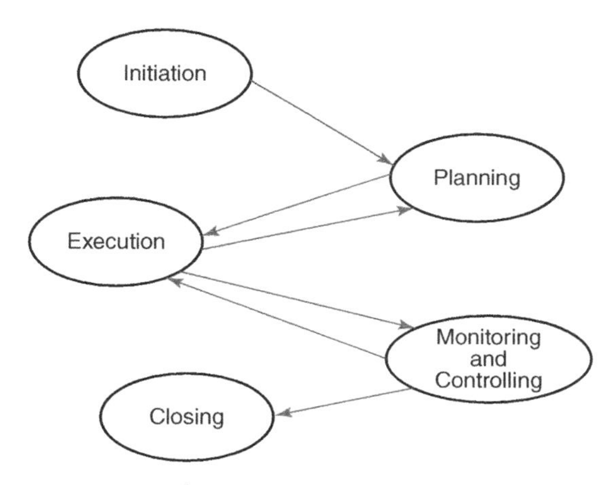

Figure 3.1 Interactions among Project Process Groups

It is the project manager's responsibility to determine how the process groups will be applied for each project. For some projects, detailed management plans will be required. For other, simpler projects, the project manager and project team may only need to give scant attention to some of them. Planning components can also be formal and written, or they can be informal and verbal. Once an organization reaches a higher degree of project management maturity, it should have developed reusable planning documents. This book provides some templates for projects that can be reused and simply modified for new projects.

These process groups can be thought of as necessary steps for project management from start to finish. In addition to applying these process groups to the project as a whole, we can also apply them to project phases. The five process groups are detailed as follows.

Project Initiation

Project initiation will be explored in more detail later during the discussion of project integration (one of the project management knowledge areas) and the primary tool used for it—the project charter. The goal of project initiation is simply to start the project or a project phase in an orderly manner, with as much agreement and understanding as can be mustered about it. At the close of project initiation, the project should have been authorized by the organization so that the assets of the organization can be employed for the project. This is particularly important in the public sector, where staff can find themselves in big trouble for initiating a project and consuming public resources without all of the necessary authorizations, which may be required from multiple authorities. At the point of project or phase initiation, complete details of the project are often not known. Those details will come later as the project progresses.

Sometimes, activities are undertaken before initiation to identify the projects to pursue or to determine their feasibility. For example, before the initiation of a systems development project, we might engage in an effort to determine if we ought to buy prepackaged software or develop software using organizational resources. In other cases, the analysis of options can be an early phase of the project rather than a preproject activity (or a separate project).

Careful project initiation is especially necessary in the public sector, because projects there can come at us from all directions, sometimes without the full sanction of the organization. Some are embedded in legal requirements, but others may simply arise from the latest idea that a deputy director had. In the private sector, processes are often in place requiring more careful analysis of cross-impacts and other opportunities before a project is embarked on. Frequently, the project's return on investment (ROI) must be calculated before initiation. The public sector is less careful with resource utilization at the departmental level, though exceptions clearly exist. (In the public sector, financial controls exist at the budgetary level in order to ensure that broad categories of costs fall within appropriated amounts and within the program definitions that are identified in the budget.) One of the goals of project initiation is to ensure that the organization understands the project and is fully supportive of it (this will be discussed later).

Even in the private sector, most people and organizations do not do a very good job of evaluating alternatives before initiating a project. Too often, people do not take the time to consider other ways to spend their time and money before working on the project that is currently in process. Because of that lack of project evaluation and prioritization, projects are simply added to the "pile" of projects, and, because staff persons often assume that management's highest priority must be the project most recently added to the pile, they engage in project "composting." That phenomenon occurs when projects are not finished or closed and, over time, they just disappear down at the bottom of the pile. (I once asked a department in a large organization how many projects were currently ongoing. The answer was "90," a number of projects that could not have possibly been managed or supported in an organization of that size. To make matters worse, very little could be identified about the status of most of them. A majority of those projects, in my terminology, had been composted.)

Most organizations would get better results if they better assessed all of their options and took the time to make sure that there is agreement about the project and a good understanding of it before getting started.

Planning

Once a project is underway through good initiation, the planning process group can begin. In that process group, the project plan, which includes the project or phase scope, schedule, budget, risk management plan, and communications plan, is created. Other elements of the project plan will be discussed later.

It is obvious that a balance needs to be reached between too much planning and not enough. For a two-week project, it is probably not worthwhile to spend a week planning it. But, most of the time, we are better off if we spend more time planning than we are used to. There is an adage in project management that says that we have the choice of "sweating the planning" or "bleeding the execution." The implications of the adage are clear: We are better off if we spend an adequate amount of time planning for the project and identifying project requirements. That implies that if we plan well, problems can be avoided down the road.

As we create the plan, we will also have to push back against managers with little patience who may want us to "just do it" instead of spending more time and resources on project planning. We will have to work to convince them that planning is a good investment.

Execution

It may seem that we spend an inordinate amount of time on project planning, but we have to have that plan in place before project execution. That is because the project manager "manages to the plan." The plan, in other words, is the blueprint for the work, and the job of the project manager is to ensure that the plan gets implemented. In project execution, the project manager assigns work to others or, depending on the size of the team, gets work done himself or herself. The major output of execution is the creation of work results.

As we get into project execution, we may (and nearly always do) discover that our plan is lacking in one way or another. If the work we are doing or need to do does not match the plan, we have two choices: (1) we can revise the work we are doing so that it conforms to the plan, or (2) we can back up and revise the plan. In most complex projects, we do both.

Monitoring and Control

When project execution is completed, we have work results. That is, we have gotten things done. But we do not know yet if they are the right things, if they conform to our quality standards, if they meet the needs of our project, or if they match our project plan. In order to evaluate our work results, we need to enter into our next process group—project monitoring and control.

Monitoring and control performs a simple function. It compares the work results created to the plan that was built. If the work results conform to the plan, we are happy. If they do not, we have two choices. If our project did not create what our plan intended, we can back up to project execution and redo the work. That is called rework and can take time and money, sometimes time and money that are not available.

Two other functions occur during project monitoring and control—managing project changes and ensuring the acceptance of our work. In public-sector projects, changes always occur. Although we might create a great plan driven by a lot of good thought about the project, changes will start happening almost as soon as the ink is dry. The job of the project manager is to prevent *unnecessary* changes, but some of the changes that occur will be necessary and good for the project. In many public-sector projects, the ultimate goal of the project may change as the project progresses. Although a project might have a broad set of goals and intended policy impacts, budget and schedule constraints can require that more limited goals be adopted.

The job of the project manager, when necessary changes come along, is to make sure that those changes are implemented and that the project plan is changed to reflect them. Often, a change in one area of the project will ripple throughout the project. For example, if a team member is assigned other work that is deemed to be a higher priority than our project, which frequently happens with public-sector projects, and, as a result, is late delivering a key deliverable:

- The risk of the entire project being late will go up.
- Additional resources may need to be assigned to that deliverable.
- Other deliverables may need to be deleted in order to stay on schedule.
- Costs may increase if more resources are allocated for that deliverable.
- The quality of the project may decline.

Making the right trade-offs between the stability of the project plan and the flexibility necessary to take emerging conditions into account is one of the project manager's most important challenges.

Last, once work results are created and it is concluded that they match the results intended in the plan, the stakeholders need to evaluate the results. That is called scope verification. Verification is performed at the close of a project or a project phase to ensure that products of use to customers have been created and to determine whether the project should be closed or proceed to the next phase.

There is a difference between the *acceptance* of work and the *correctness* of the work as measured by quality standards. For example, for a public systems development project, the user community may accept the system and indicate that it meets their minimum requirements (acceptance). It is quite another thing for that system to meet the exacting quality requirements set for it (the correctness of work according to standards). Correctness of deliverables is a measure applied to project quality control and will be described later.

Closing

When the project is over, it needs to be brought to a close. The purposes of closing projects are to:

- Document the results and create the project archives
- Identify lessons learned so that the next project can be managed even better
- Celebrate success with the project team and stakeholders
- Send project team members off to their next projects with good feelings

Closing the project takes time and attention. Unfortunately, often the effort is not made, and projects remain in an ambiguous state in which no activity is performed but no clear closing has been made. Too often, work just stops on projects without their being closed (and, as described above, they compost).

Asking how many projects are currently open is an interesting and worthwhile exercise. The predominant answer is likely to be "we are not sure." If a count can be made, it is likely that the count will exceed the number of projects that the agency can realistically hope to complete in any reasonable time period. Until a disciplined approach to project management is adopted, which includes project closing, project inventory cannot be managed.

PROJECT MANAGEMENT KNOWLEDGE AREAS

The *PMBOK® Guide*—Fourth Edition identifies nine project management Knowledge Areas. These are areas that the project manager and the project team must be concerned with. They include:[1]

- Project Integration Management
- Project Scope Management
- Project Time Management
- Project Cost Management
- Project Quality Management
- Project Human Resource Management
- Project Communications Management
- Project Risk Management
- Project Procurement Management

These Knowledge Areas and their implications for public-sector projects will be described in the chapters that follow.

THE TRIPLE-CONSTRAINT MODEL

The triple-constraint model of project management simply implies that there must be a balance in every project between (1) the project's scope, (2) the time available for the project, and (3) the resources available for it.

---

[1] Project Management Institute, *A Guide to the Project Management Body of Knowledge (PMBOK® Guide)*—Fourth Edition, 2008, p. 43.

Initially, it is the project team's goal to create a balance among those elements to create a project that can be successfully managed. Later, if one of those elements is adjusted, one or more of the other elements will have to be adjusted as well.

For example, in a business development project of the type undertaken by government departments of development to recruit business to a jurisdiction, a key stakeholder may decide midway through the project that the project team needs to explore the availability of federal grants for which the company may be eligible. If that deliverable (and the activities that go with it) are to be added, more time or more resources or both are needed. The only other option is to reduce scope somewhere else in the project.

Recall that adding scope without recognizing the impacts of it is a symptom of magical thinking, which can be eliminated through good project management. The same is true of reducing project resources or moving deadline dates up.

The triple-constraint model also introduces conflict into projects. If scope is added to a project, some stakeholders will favor adding additional resources to ensure balance. Others might want to add more time to the project. Still others might favor reducing some other portion of the project scope. It is the job of the project manager to build consensus, resolve conflict, and manage stakeholder expectations around the tripe constraint model. The critical role of project conflict management will be described in more detail later.

PROJECT PROCESSES

The *PMBOK® Guide*—Fourth Edition, links the five Process Groups of the IPECC model (using the extra M for monitoring) with the nine Knowledge Areas to create a five-by-nine matrix. In the cells of that matrix, the *PMBOK® Guide*—Fourth Edition, identifies 42 processes. (Not all cells in the matrix contain processes, and some contain more than one.) One of the challenges of the project manager is to identify which of those 42 processes should be applied to each project or project type.

Instead of applying that set of processes, a set of required functions is posited here for meeting the unique constraints and challenges of public-sector projects. Those functions are listed in Tables 3.2 through 3.6. The processes identified in the *PMBOK® Guide*—Fourth Edition, which correspond to these required functions for public-sector project management, are included in the third column of the tables. Although we could have

Table 3.2 Project Initiation

Knowledge Area	Public-Sector Project Functions	*PMBOK® Guide*—Fourth Edition Processes[a]
Project Integration Management	Framing and initiating the project	Develop Project Charter
Project Communications Management		Identify Stakeholders

[a]Project Management Institute, *A Guide to the Project Management Body of Knowledge (PMBOK® Guide)*—Fourth Edition, 2008, p. 43.

Table 3.3 Project Planning

Knowledge Area	Public-Sector Project Functions	*PMBOK® Guide*—Fourth Edition Processes[a]
Project Integration Management	Developing the project plan	Develop Project Management Plan
Project Scope Management	Creating a scope management plan Defining and confirming project scope	Collect Requirements Define Scope Create WBS
Project Time Management	Defining project activities Creating the project schedule	Define Activities Sequence Activities Estimate Activity Resources Estimate Activity Durations Develop Schedule
Project Cost Management	Creating the project budget	Estimate Costs Determine Budget
Project Quality Management	Identifying project quality requirements	Plan Quality
Project Human Resource Management	Creating a plan for optimizing the use of human resources	Develop Human Resource Plan
Project Communications Management	Creating a communications plan	Plan Communications
Project Risk Management	Creating a plan for risk management	Plan Risk Management

(*continued*)

Table 3.3 (*continued*)

Knowledge Area	**Public-Sector Project Functions**	***PMBOK® Guide*—Fourth Edition Processes[a]**
	Identifying risks Analyzing risks Developing risk responses Creating a plan for managing the project's legal and administrative constraints	Identify Risks Perform Qualitative Risk Analysis Perform Quantitative Risk Analysis Plan Risk Responses
Project Procurement Management	Identifying necessary purchases and resource acquisition Working with purchasing offices to identify and select vendors	Plan Procurements

[a]Project Management Institute, *A Guide to the Project Management Body of Knowledge (PMBOK® Guide)*—Fourth Edition, 2008, p. 43.

examined the processes identified by the *PMBOK® Guide*—Fourth Edition, we have focused instead on the functions listed as follows because of the differences or special challenges they pose for public-sector project management.

Table 3.4 Project Execution

Knowledge Area	**Public-Sector Project Functions**	***PMBOK® Guide*—Fourth Edition Processes[a]**
Project Integration Management	Performing project work	Direct and Manage Project Execution
Project Cost Management	Acquiring the financial resources for the project	
Project Quality Management	Managing project quality	Perform Quality Assurance
Project Human Resource Management	Motivating and managing the project team Resolving project conflict	Acquire Project Team Develop Project Team Manage Project Team
Project Communications Management	Providing information to project stakeholders	Distribute Information Manage Stakeholder Expectations
Project Procurement Management	Managing contracts and vendors	Conduct Procurements

[a]Project Management Institute, *A Guide to the Project Management Body of Knowledge (PMBOK® Guide)*—Fourth Edition, 2008, p. 43.

Table 3.5 Project Monitoring and Control

Knowledge Area	**Public-Sector Project Functions**	***PMBOK® Guide*—Fourth Edition Processes**[a]
Project Integration Management	Comparing the work to the plan and managing changes	Monitor and Control Project Work Perform Integrated Change Control
Project Scope Management	Controlling scope Verifying acceptance of deliverables	Control Scope Verify Scope
Project Time Management		Control Schedule
Project Cost Management	Managing project costs and reporting on project expenditures	Control Costs
Project Quality Management		Perform Quality Control
Project Human Resource Management		
Project Communications Management		Report Performance
Project Risk Management		Monitor and Control Risks
Project Procurement Management		Administer Procurements

[a]Project Management Institute, *A Guide to the Project Management Body of Knowledge (PMBOK® Guide)*—Fourth Edition, 2008, p. 43.

Table 3.6 Project Closing

Knowledge Area	**Public-Sector Project Functions**	***PMBOK® Guide*—Fourth Edition Processes**[a]
Project Integration Management	Closing the project and contracts	Close Project or Phase
Project Communications Management	Capturing and managing knowledge	
Project Procurement Management		Close Procurements

[a]Project Management Institute, *A Guide to the Project Management Body of Knowledge (PMBOK® Guide)*—Fourth Edition, 2008, p. 43.

APPLYING PROJECT FUNCTIONS AND PROCESSES FOR PUBLIC-SECTOR PROJECTS

Choosing the right mix of functions and processes from the list and the degree to which they should be applied for a public-sector project may seem like a daunting task. Indeed, it is unlikely that any project would employ all of the listed functions and processes. Fortunately, guidance can be provided to assist public-sector project managers who are attempting to make that choice. That guidance follows. Successive chapters examine the required functions and the documents and tools they employ in more detail.

It is also important to remember that these functions and processes can be applied with different levels of rigor. For example, on a small project, we might just create a short charter within an e-mail and circulate that. On a longer, more complex project, we might create a detailed project charter with extensive detail and routing.

The following steps encapsulate the project management functions that are described later. They are presented here to provide context for that later discussion and an overview.

Step 1: Get the Project Started Correctly

Nearly every project can benefit from a good initiation process. Public-sector projects appear from all directions. Some are enabled by statute, some are embedded in administrative rules, some arise from citizen demands, and others are identified by public-sector managers. Very few come with enough detail to get the project moving in the right direction, however. As a result, the project manager has to ensure that every project is initiated with some degree of rigor. No matter what documents or formats the process of initiation employs, the project manager has to ensure that:

- The organization has authorized the expenditure of public resources on the project.
- The principle stakeholders agree on the general direction of the project.
- The project manager and the project team understand what is being requested and agree to dedicate their best efforts to the project.

Project initiation, typically using a project charter, creates a dialogue about the project that helps define it and creates a document that can be used for later reference. The charter represents an agreement between the project team and the organization. In the charter, the organization identifies what it wants to have done, and the team indicates the circumstances under which the results can be delivered.

The purposes of public-sector projects often are not well stated at their initiation. That requires project teams to interpret the project on their own, which is a high-risk behavior. Any public-sector project manager who has gone down the wrong path on a project can painfully detail the consequences that occurred when their error was detected, and, unfortunately, the public sector has highly developed processes for assigning blame and identifying scapegoats.

A better strategy than trying to guess what stakeholders expect is for the project team to build a charter document—if one has not already been prepared—and circulate that document to the necessary stakeholders. Essentially, the project team is saying to the organization, "Here is how we interpret this project. Let us know if you see it otherwise." At the beginning of a project, when the charter is prepared, little is known about the project, and progressive elaboration has just begun. The goal of initiation is to identify what is known about the project and codify it as a starting point.

In some organizations, the project charter is created by someone outside the project team. That is, the project may be handed to the team from some customer organization (e.g., the legislature, an internal department needing services, or some other source) or a senior manager. That customer or senior manager could be required to complete the charter before the team will take on the work. In those cases, the project team can later continue the progressive elaboration process by adding details to the charter as project planning progresses.

When public-sector projects are initiated, the sponsor or project team should take the time to identify the constraints that are likely to affect the project. These constraints can include such items as legal constraints and the requirements to obtain the approval of oversight agencies. In public-sector projects, those constraints can significantly impact the project and even stop it in its tracks.

Step 2: Build a Triple-Constraint Model

Once the project has been authorized to expend resources, the project team can begin the planning process. That planning process should begin with construction of the triple-constraint model for the project. As was detailed earlier, the triple-constraint model posits that a balance must be struck among the project's scope (what it will do), its schedule, and its cost. The project plan must begin with the identification of the project's scope.

Scope

Scope planning creates a plan for managing project scope, including processes for documenting and changing the scope. Scope definition continues the progressive elaboration process and creates documents necessary for identifying and building consensus on the project's scope.

Creating the work breakdown structure (WBS) is the most important process in project planning. It expands the list of deliverables identified in the charter or scope statement to include all of the deliverables necessary for successful delivery of the project. In the WBS, they are arrayed in a graphical or outline format. The WBS and its associated WBS dictionary define the full scope of the project. If a deliverable is on the WBS, it is part of the project. If it is not, it is not. No public-sector project should be attempted without a good WBS. Some would even assert that ensuring that a WBS is in place is a matter of professional responsibility for the project manager rather than simply a good best practice.

Time

We can define the time necessary for completing the project (and create that side of the triple-constraint model) by creating a project schedule. First, the activities necessary to complete deliverables must be identified. Then, they have to be sequenced, the durations estimated, the resources necessary for their completion identified, and the schedule for the project created. The project team may discover that the time required for completing all of the deliverables exceeds the time available. They must then engage in schedule compression.

The unique constraints of public-sector projects and processes usually have a profound impact on the project schedule. These constraints and the time it takes to cope with them have to be factored into the project schedule.

Cost

The third side of the triple-constraint model is cost. To build that side, we need to estimate the costs of the project and reconcile those costs to the budget cycle. In some public-sector projects, detailed cost estimates are not performed. Too often, public-sector managers presume that staff on the payroll is free for projects. At a minimum, some cost proxies need to be applied so that the relative level of project effort can be identified. We need to know if the project will engage the entire department for an extended period or if the sponsor imagines the project as a relatively small initiative that will only take the time of one or two people for a short period.

Instead of identifying detailed costs, some organizations use the number of hours of work assigned to the project as that proxy indicator of effort.

Step 3: Apply other Planning Processes as Necessary

With the triple-constraint model as the basis of project planning efforts, we can now examine the other planning processes to determine which we should apply given the size and risk of our project. One set of planning processes that leap out for special consideration are those related to risk planning. We need to engage in risk planning to identify risks and develop plans for managing them. Those risk management plans often create additional project costs and activities, which affects the budget and schedule.

Not surprisingly, public-sector projects have a substantial and unique set of project risks no matter how large or small the project. Being aware of those risks and planning as well as possible for them is critical to good project management. In addition, public-sector projects are constrained in multiple ways. Managing those constraints can take considerable time. We may also want to create plans for identifying project quality requirements, optimizing the use of human resources, identifying necessary purchases and resource acquisition, and creating a plan for managing the purchasing and resource acquisition processes.

Step 4: Integrate the Planning Documents Created and Build the Project Plan

The project management plan is the combination of all of the planning documents. To a simplistic degree, the project management plan can be thought of as a stapled compendium of the other documents. In reality, those planning elements have to be integrated to create a complete plan, and consensus must be developed among stakeholders on that plan.

Step 5: Do the Work of the Project

In the process group called project execution, the work of the project is done. More directly, the work identified in the plan is completed. Other processes in the project execution process group can be applied as needed. To accomplish the work of the project, the project team must be motivated and managed, which can be a sizable challenge in public-sector projects, given the lack of performance incentives and penalties available to public-sector project managers.

Step 6: Compare the Work Completed to the Plan, Make Changes as Necessary, and Ask Stakeholders to Accept the Work

In the monitoring and controlling process group, the work of the project is compared to the work that was planned. If the work does not match the planned work, the work can be redone. The plan can also be changed if that is appropriate. Changes have to be integrated into the project plan and plans for work adjusted as necessary. Deviations between the work and the plan can occur with regard to project scope, time, cost, quality, risk, human resources, risks, and the work of vendors.

Last, in the monitoring and controlling process group, the work products of the project must be presented to its customers. That occurs when stakeholders are asked to verify the acceptance of deliverables. If they do not accept the work, the project must be cycled back through the process groups. If they do accept the work, the project can be closed. Any changes in the project require careful integration across all the elements of the project plan.

Step 7: Close the Project

The project should be formally closed, though few public-sector organizations engage in solid project-closing activities. Closing activities include identifying lessons learned, archiving project documents, celebrating successes, and sending project team members off to their next assignments.

THE NECESSARY SKILLS FOR PUBLIC-SECTOR PROJECT MANAGERS

Project managers in either the private sector or the public sector require a broad array of skills in order to manage their projects in complex business environments. Some of those skill sets include:

- Hard skills, like scheduling, budgeting, performance management, and quality control
- Soft skills, like team building, conflict management, negotiations, motivating team members, and managing stakeholder expectations

Project managers in the public sector, however, require a more sophisticated skill set that also includes:

- Understanding of government processes
- Political awareness and sensitivity (Though project managers in the private sector are required to understand organizational politics, they do not operate in an environment that so frequently contains a direct opposition party that measures its success and ability to get elected, in part, based on the failure of the party in power.)
- Managing employees without the ability to provide the types of incentives (and disincentives) available to the public sector
- Operating among a vast array of stakeholders, including the public, administration officials, and legislatures
- Understanding the press and operating in an environment of high visibility with little organizational tolerance for failure and media ready to exploit that failure
- Managing conflict with internal and external stakeholders to an extent beyond that required of private-sector project managers

As a result, public-sector project managers have to be more adept at managing the context of the project and its environment than most private-sector project managers. They have to have the ability to focus on the project and its internal management *and* the external environment, which may be hostile to the project. (Remember that only the public sector has built-in opposition in the form of the "other" party.)

As noted earlier, public-sector project managers need to develop the habit of sharing successes and not placing blame for failures. They need to be able to work with disparate groups, maintain their perspective and a sense of humor, recover quickly from adversity, manage conflict, and be relentless in their pursuit of successful projects but diplomatic and sensitive to the needs of others.

Unfortunately, these skill sets are more difficult to transfer than the more concrete project management skills. Later, some methods by which public-sector project managers can manage the complexity and chaos that can accompany public projects will be identified. Management of project uncertainty requires relentless communication and constant attention to and involvement of project stakeholders.

DISCUSSION QUESTIONS

1. What types of public-sector projects have you been engaged in? What others can you imagine?
2. What experiences have you had with a PMO? Do you think a PMO would be of use to your agency? What would its mission be?
3. In what ways do your projects require trade-offs to be made? What happens when someone demands an increase in scope without a corresponding increase in resources or time? What options do you have for responding?
4. What skills do you think project managers in the public sector require? What skills are most important? How can those skills be learned or transferred?

EXERCISES

1. For a project you are familiar with, break it down into logical phases. What are the deliverables of each phase?
2. For a project you are familiar with, identify which of the processes listed that you would apply. Why would you apply those?
3. For the Apollo Project described earlier, identify some deliverables and some activities that would be undertaken to create them. Make sure that you use nouns for the deliverables and verbs for the activities.

The Marshall Plan

Some public-sector projects have grand objectives but require more subtle implementation and deployment than building a burial chamber or launching a man to the moon. Some require the navigation of tricky political currents and forces.

At the close of World War II, Europe was in ruins. Steady bombing and the land war that had been fought as the Allies converged on Berlin had reduced major cities to rubble; the transportation infrastructure had been decimated, agricultural devastation threatened famine, and entire economies had been brought to a near standstill. Most of the nations of Europe were financially strapped as a result of the massive amounts of money spent on the war.

Secretary of State George C. Marshall argued that America should do whatever it could to assist those nations in their return to economic health. The United States was aware that similar conditions had followed World War I and had led, in part, to the pressures that had later fueled World War II. Although the Allies had hoped that Europe would recover on its own after World War II, by 1947, they had seen little improvement, and resource shortages were bringing the problems to a head. On humanitarian and political grounds, the United States knew that Europe had to be helped to its feet as quickly as possible, in part, to head off growing Soviet influence in the area. In fact, the supporters of the Marshall Plan were handed a strong argument for the establishment of the plan when a Soviet-backed, Communist coup took place in Czechoslovakia.

In the United States, two options for rebuilding Europe were discussed. One required that money be taken from Germany in the form of war reparations. The other plan, pushed by Marshall—who believed that a strong Germany was vital to European stability and economic growth—required replication in Europe of the New Deal programs believed by some to have rescued the United States from the Great Depression.

Marshall's strategy required the European nations to build their own plan for economic recovery, which the United States would fund, and the plan ultimately became a joint American–European endeavor. Sixteen nations ultimately met in Paris to develop that plan, which strongly emphasized free trade. After much debate and compromise, that plan was approved by the U.S. Congress, and, in

1949, the Marshall Plan began the movement of what was to finally total $12.4 billion in aid to Europe.

Despite its challenges, the Marshall Plan accomplished its goals. By the time it ended in 1953, starvation was no longer threatening Europe, political stability had been achieved, a period of economic growth had begun, and the Communist influence on Western Europe had been greatly reduced. Some also argue that the Marshall Plan paved the way for the economic integration that resulted in the formation of the European Economic Community. It clearly met the demand for a combination of assistance and self-help by the nations of Europe and provided a practical program while allowing the United States to be generous and idealistic.

Chapter 4

Project Integration

PUBLIC-SECTOR PROJECT INTEGRATION: WRESTLING WITH THE OCTOPUS

This chapter begins an exploration of the project management knowledge areas. Each of the nine knowledge areas identified by the *PMBOK®* *Guide*—Fourth Edition will be described, as well as the particular implications for them in public-sector projects.

Project integration is the set of project management activities and processes necessary for coordinating and combining the elements of a project. It allows the project manager, project team, and project stakeholders to fit the various elements of the project together to make sure that cross-impacts are managed to create results that satisfy all of the stakeholders. This section examines each of the required functions for public-sector project integration management.

Even in the best conditions, project integration can be a challenge. For example, the strategies that the project team develops for managing its risks can create additional activities that need to be performed and additional costs. A project schedule change can increase risks and cause other project impacts. Scope changes always cause ripples through the project plan.

Because of the complexity of public-sector projects, project integration in that sector can take on additional complexity. Public-sector projects rarely involve isolated parts of an organization, but rather frequently reach, octopus-like, into multiple organizations and impact multiple stakeholders. As mentioned earlier, public-sector projects are particularly susceptible to the impacts of those stakeholders, such as oversight agencies and legislators, who can influence the outcome of a project. Although private-sector projects can engage a variety of stakeholders who have to be satisfied, public-sector projects are impacted by a broader variety of stakeholders who can stop the project, challenge it publicly, or influence how it is conducted.

For example, in a small municipality, trash collection services were outsourced from a city department employing public employees to a private vendor. The project, in concept, was relatively simple. Although it appeared that the city officials had the authority to enter into that contract, some residents objected to the decision and attempted to have it overturned. Stakeholders for that project initially were limited but soon expanded to include:

- The court system, including the State Supreme court, which eventually ruled against the ballot initiative
- Citizen groups
- Attorneys
- The Secretary of State, who was responsible for crafting ballot language
- The media
- Collectors of signatures for the ballot initiative
- The contract company and the former city employees
- Voters

In the end, the actions of the city council were upheld, and the contract stayed in place, though the costs of the project presumably were increased due to the extra efforts required to manage legal challenges and work with stakeholders.

Those stakeholders set project constraints and additional project requirements. As a result of the influence of these types of stakeholders, public-sector project managers can quickly become drawn into larger issues of public policy than first anticipated and into engagement with broader groups who have an ability to affect the project. The initial focus of the project—in the example, a focus on purchasing options and the delivery of trash collection services—can quickly be expanded and cause public-sector project managers to be required to apply expertise far outside their primary area of knowledge.

For any project, a communications plan should be created that identifies stakeholders and their needs and determines the best way to communicate with them. That plan can be formal or informal, depending on the complexity and needs of the project, and will be described in more detail in a later chapter. In the public sector, however, project managers have the extra challenge of integrating a broader array of project influencers as well.

Those influencers for even a relatively simple public-sector project may include:

- Minority parties and the political opposition
- Legislators who can change the political and legal landscape
- The media
- The courts
- Senior administrative agencies that control purchasing and hiring
- Advocates, whose interests may run counter to those of the project or who might be relied on for support
- Policy makers in other agencies who might have projects in process that require integration
- Audit officials who set rules for processes and transactions

Project integration in the public sector, therefore, becomes a process of integrating the varied components of the project (the internal complexities) and the varied influencers of the project (the external complexities).

OVERVIEW OF THE NECESSARY FUNCTIONS FOR PUBLIC-SECTOR PROJECT INTEGRATION

Project integration consists of the activities and processes necessary for getting the project started, pulling together all of the varied elements of the project into a consistent plan, performing the work of the project, monitoring the work, integrating changes, and closing the project. Project integration can be applied to either a project or a project phase. Although the required activities can be defined in a linear fashion, they are often looped or iterated as the project progresses. For example, project execution for a project designed to attract a new business to the state may be well underway when new legislation takes effect that provides additional incentives for new business. That will require, at a minimum, a change in the project plan to incorporate the exploration of the new options.

The required functions for public-sector project integration are:

- Framing and initiating the project
- Developing the project plan
- Performing the project work
- Comparing the work to the project plan and managing changes
- Closing the project and contracts

Each of these functions is described in brief, and then the development of the project charter is discussed in more detail.

Framing and Initiating the Project

Framing and initiating the project requires getting the project started correctly, with a consensus on what the project is about. It also requires identifying what is known about the project at this point and getting approval from authorized managers for the project.

As noted earlier, the purposes of framing and initiating the project are to ensure that:

- The organization has authorized the expenditure of public resources on the project.
- The principle stakeholders agree on the general direction of the project.
- The project manager and the project team understand what is being requested and agree to dedicate their best efforts to the project.

In public-sector projects, framing the project may include interacting with stakeholders to build early support for it and to identify opposition that may need to be factored into project plans and the scope of the project. It may also include development of a business case for the project in order to prove its value and compete with other projects to gain budget authorization. At this point in the project, the project manager and sponsor may have to take on the role of selling the project to stakeholders. The most adept public-sector project managers understand that they need to give stakeholders the opportunity to co-create the project with them. Rarely can public-sector project managers or sponsors of the project create a project unilaterally without the participation of a broad set of stakeholders. The best public-sector project managers have an appreciation for the give-and-take that surround any activity in the public sector and know when to hold firm and when to acquiesce to the needs and opinions of stakeholders.

Once the project has been vetted and is certain of support, the project charter can be developed. The project charter is the initiating document of the project. It authorizes spending the organization's assets on the project. The purpose of the charter is to create as much understanding as possible at the start about the purposes of the project and expectations for its performance.

Because the charter is the initial document of the project, there are no other project documents to build on. However, organizational assets, such as templates, standard forms, and a project management methodology, may help with managing the project. The environment of the project and its assets also need to be considered to help get the project started.

Some public-sector organizations have a standard charter document that is prepared by someone needing the project's results. For example, in some information technology offices, a customer or user is required to file a request for assistance or a system enhancement. However, because of the wide diversity of public-sector projects, most public-sector organizations do not have a standard project charter document and find projects arising from a variety of sources, including legislation, management strategies, and public requests.

By the time the charter document is completed, the project manager should have been selected and assigned. In some cases, the development of the project charter is a simple exercise. In other cases, it will require substantial dialogue and negotiation if key stakeholders have differing views. Beginning a project without a clear idea of what the project is supposed to do, as expressed in the charter, can be a disaster waiting to happen.

If no project charter is put in place and agreed to by the major stakeholders, the project manager and the project team are placed in a risky situation. If agreement on the charter has not been achieved, later in the project, stakeholders can:

- Allege that they never supported the project
- Argue against the dedication of resources to the project
- Create their own versions of the project's outcomes
- Allege that the project team is operating without organizational authority
- Raise objections that were not made clear at the outset

- Object to the project's organization or approach to the problem
- Assign resources to higher-priority projects instead of making good on commitments to the existing project
- Demand performance from the project team that is not possible under the constraints the project faces
- Revise project assumptions

Some project charters contain a lot of details, whereas others are relatively simple. In its most simple format, a project charter should contain:

- *The business need for the project.* This need should be stated as a problem to be solved or an opportunity to be pursued. For public-sector IT system deployment, "system replacement" is not a business need. The system is being replaced for a reason, like better responsiveness for users or the continuation of system support.
- *The solution that this project will apply to meet the business need.* For any business need, there are a host of solutions, only one (or some) of which is to be applied by this project. For example, a government agency may identify that high rates of employee turnover in key areas is damaging its ability to accomplish its mission. Turnover can be addressed by increasing pay, improving benefits, providing more flexibility, offering training, improving morale by having more social events, and other options. The options that this project intends to pursue are identified in the charter.
- *Assumptions.* Assumptions are those uncertain outcomes that, for planning purposes, are assumed to be true. For example, we might assume that the necessary resources for the project will be provided. That might not be true, but we may not be able to manage the project without those resources. As a result, we list that assumption in the project documents that we circulate to stakeholders. By stating that we will get the necessary resources as an assumption, we are serving notice to the stakeholders that we expect to have those resources. Stakeholders might review these assumptions and tell us that they are unreasonable. If they do, we have to rethink the project.
- *Constraints.* Constraints are any factors that limit the project team. As described earlier, public-sector projects are severely constrained. Normal project constraints, like time limits and a budget, impact all projects.

- *Overview of the legal and administrative context for the project.* For public-sector projects, the charter should also include an overview of the legal and administrative factors that the project must deal with. These requirements will impact the project plan in substantial ways. The charter could include a listing of oversight agencies whose dictates must be complied with, laws and rules that apply to the project, and stakeholders who must be integrated into the project in order to ensure their sign-off on the project's compliance with rules and policies.

Project charter documents can include other elements that are known about the project at the time of its initiation. Of course, any elements contained in a project charter are described at a very high level of generalization, because the progressive elaboration has barely been started.

As indicated earlier, someone other than the project manager may have prepared the project charter, and the project manager may have been appointed after the charter has been prepared and submitted to the organization responsible for the project. If that is the case, it is the project team's first chance to elaborate the details of the project and put their stamp on it.

A second document is sometimes used in project initiation. That document is sometimes called the scope statement or preliminary scope statement. Some organizations also simply continue to add details to the project charter. Those additional details could include:

- A more detailed definition of the scope of the project, including the identification of very high-level deliverables
- The composition of the project team
- Stakeholder requirements
- Other involved organizations
- The requirements for project reporting
- Identification of laws and rules that apply to the project
- High-level risks
- Cost estimates
- A tentative project schedule, including summary milestones

Like the project charter, this additional detail has to be circulated to the appropriate stakeholders. In addition to creating building blocks for the management of the project, another goal of this additional detail is to continue to build agreement about the project and to manage the expectations of stakeholders.

Sometimes, a project arrives without a charter. It might just be an idea that the agency director had or something that was required by statute. If a project arrives without a charter, the best strategy is for the project manager to write one. Once it is written, the charter can be circulated to stakeholders, essentially asking, "This is how we see this project. Do we have it right? Do we have your authorization to proceed in this manner?" Most executives respond very positively to an effort by the project team to discover what, exactly, the executive needs.

A template, including the mandatory elements and a few other useful details, for a public-sector project charter follows in Table 4.1.

It is important to remember that the project charter is created through dialogue with the project stakeholders. A project charter that only the project team is aware of or agrees with is worthless. A charter that has been agreed to by all of the relevant stakeholders will serve the project well and can provide a solid basis for later project decision making.

Table 4.1 Project Charter Template

Project Name	
Need for the project	Describe the need for the project, which is often described as the problem that the project intends to solve.
Description of the solution	At a high level, describe the solution being used.
Project schedule milestones	Identify the most significant schedule milestones for the project.
High-level deliverables	Identify the major deliverables for the project. Deliverables are nouns.
Stakeholders	Identify those people or groups who are affected by the project, involved in it, or who can influence the outcome.
Assumptions	Identify those things assumed to be real or certain for the project.
Constraints	Identify those factors that limit the project.
Legal and administrative factors	Identify those laws, rules, processes, and oversight agencies that must be dealt with and the amounts of time and resources likely to be required.
Assets	Identify the assets available, including a list of those who can help and personal assets.

Developing the Project Plan

As you may notice, the next several required functions for public-sector project integration align very closely with the process groups of the IPECC model. The first is the development of the project plan. The project plan, sometimes called a project management plan, is the compilation of the other elements of project planning (e.g., the budget and the schedule) into a coordinated document that lays out the intentions of stakeholders and the means the project team has selected to accomplish those intentions. Building that plan is an exercise in negotiations, coordination, and flexibility.

The project plan is not a static document, nor is making changes to the plan an indication of the project manager's inability to effectively plan. As we undergo progressive elaboration of the project, we are always learning new things and adapting our efforts. Those changes should be reflected in the plan as they are made. Some project managers, in fact, become unnaturally wedded to their plans and deploy them no matter what. That is a mark of a lack of adaptability and will probably result in project results that do not meet the needs of stakeholders.

How much project planning should be undertaken? There is no firm answer, but it is likely that we do not do enough planning for public-sector projects before we launch them. One of the adages of project management is that we can either "sweat the plan" or "bleed the execution." In general, time spent on careful project planning and building consensus about the project is time well-spent. The extent to which we plan for our projects should be determined by the level of their risk.

Elements that should be included in the project plan include:

- Project charter
- WBS
- Schedule
- Budget, including necessary integration with budget processes
- Risk management plan, with identified risks, risk analysis, and risk-response plans
- Legal and administrative constraint management plan
- Quality management plan, including quality metrics and goals
- Human resource management plan
- Communications plan
- Identification of necessary purchases and resource acquisitions
- Plan for managing the purchasing and resource acquisitions processes

These elements of the project plan will be described in those chapters of this book that address the relevant knowledge areas.

Performing the Project Work

In this process, we simply do the work of the project. More specifically, we do the work that has been identified in the project plan. If we do work that is not in the plan, we run the risk that we are doing unnecessary work, and we cannot evaluate that work by comparing it against the plan.

In this process, project managers in either the private sector or public sector are engaged in such activities as:

- Assigning duties to project staff
- Overseeing the creation of project deliverables
- Checking the quality of deliverables
- Identifying requested changes in the project, evaluating them, and modifying the project plan accordingly
- Preventing unnecessary changes
- Conducting project team meetings
- Dealing with known and emerging risks

Project execution in public-sector projects can also involve other activities. They could include:

- Working to expedite hiring and purchasing processes
- Complying with laws, rules, and guidelines
- Applying for waivers of processes and rules
- Working with legal staff to craft administrative rules and, in rare cases, proposed statutes
- Interfacing with representatives of higher-level agencies and oversight agencies
- Communicating with oversight agencies to keep them aware of the project status
- Working with legislative liaison staff to identify political opportunities and obstacles
- Completing funding and grant applications
- Integrating project plans with budgeting processes
- Providing legislative testimony
- Scheduling required public input sessions and opportunities

- Performing research on relevant rules, laws, and processes
- Engaging in progressive discipline for project team members
- Revising job descriptions or creating new job descriptions in order to allow the work of the project to be done

As most public-sector project managers can attest, some of these other activities that are either unique to the public sector or more pronounced in the public sector than in the private sector can absorb substantial amounts of the project manager's and project team's time. Although existing statutes or rules may be a given in some public-sector projects, in others, revised statutes or rules can be another project deliverable. Whether the project team has to comply with statutes and rules or is seeking to change them, additional project activities will be required and will take additional time and resources.

Comparing the Work to the Plan and Managing Changes

This required function compares the work that has been completed by the project team to the project plan and identifies differences. It can occur in the midst of execution, in which case preventive actions can be undertaken, or it can occur after deliverables have been produced, which requires corrective actions. This process brings together the more detailed activities of project monitoring and control that apply to each of the project management knowledge areas, such as controlling the schedule and controlling project costs.

Monitoring and controlling project work also involves the creation of project data that can be used to monitor the performance of the project and archived for future reference.

In public-sector projects, deviations from the project plan are often identified by stakeholders outside the project team. For example, an oversight agency may determine that budgets have been exceeded or that processes do not align with legal requirements or requirements embedded in administrative rules. The project team must, however, avoid the temptation to hide project results or impacts from those agencies. Even though working with stakeholders can be a challenge, those challenges are far easier to cope with than the challenges that result from having been caught trying to conceal outcomes. (A very wise person once told me that any problem in any public-sector project can be overcome if the problem is admitted to and help is requested.)

As changes in the project occur, those changes have to be reflected in a revised project plan. That would be a simple process of updating isolated elements of the plan when things change if it were not for the fact that project changes have the tendency to ripple through the plan. What appears to be a small change can, in fact, have many other impacts.

For example, a project designed to upgrade an existing IT system might have assumed that a sole-source contract could be entered into with a vendor who is an expert in the system. Midway through the project, the project manager may be informed by the purchasing agency that the argument made for justifying the sole-source contract is not adequate. That single change can cause:

- Schedule risks to increase (i.e., the chance that the project will be late), which may require compression of the project schedule by reducing scope, adding resources to selected tasks, or rearranging tasks to shorten the schedule
- Additional resources to be required, which may require that the budget for other activities be reduced or that additional funds be requested. If additional funds are to be requested, that, too, adds new project activities and has an impact on the schedule.
- Project quality goals to be reduced
- The need for additional communication with stakeholders in an attempt to overrule the purchasing agency
- The need for reassignment of team members, which also may have a schedule, resource, and quality impact and which may precipitate legal issues
- Additional work to identify other potential bidders
- Additional work to develop selection methods and comply with procurement processes

Although the list could go on, the point is made—integrating changes across the project is a complex endeavor that, nonetheless, is critical to maintaining the integrity of the project plan.

Activities that public-sector project managers might engage in as they compare the work to the plan and manage changes could include:

- Informing stakeholders and oversight agencies of project changes
- Clarifying alleged failures to comply with rules, laws, and policies
- Requesting exemptions from rules

- Identifying the causes of variations from the plan and taking corrective and preventive actions
- Evaluating project staff performance
- Requesting additional resources
- Engaging in change request activities for vendor contracts
- Working with procurement and legal experts to determine what actions can and should be taken with regard to contract problems or breaches
- Working with stakeholders to revise project and product requirements
- Reviewing vendor invoices and payments

Closing the Project and Contracts

No one likes to take the time to close a project. By the time a project is over:

- No time remains for closing the project.
- The budget is gone.
- Project team members are ready to move on to other challenges.
- The organization has already lost interest in this project and wants to move on to newer priorities.
- Vendors, who are not typically paid for activities required for closing, have moved on.

If we do not take the time to close projects, however, we may not be able to identify the point at which they have been completed, and we will not be able to integrate lessons learned into our project management methods and processes. As a result, we will have a harder time improving the project maturity of the organization.

Activities engaged in by the project manager and project team in closing projects might include:

- Compiling project archives, including complying with record-retention requirements and laws
- Terminating vendor contracts and authorizing final payment
- Identifying lessons learned and reporting those to the appropriate project oversight organizations
- Revising job descriptions to remove project responsibilities
- Evaluating team members
- Creating documentation of the final project outcome or product

Most public-sector organizations will be subject to requirements for archiving and retaining documents for specified periods of time. Consult an expert on the requirements that pertain to your projects and contracts.

Contracts related to the project should also be closed. Some contracts are closed before they have run their natural course (called early termination), whereas others can be deemed to have been closed when deliverables have been accepted and payment made to the vendor.

BEST PRACTICES FOR PUBLIC-SECTOR PROJECT INTEGRATION

Best practices for managing the integration of public-sector projects include:

- Work to build a culture in your organization that requires a project charter for all projects
- Engage in a process of co-creating the project with appropriate stakeholders
- Build a solid business case for the project in order to justify budget requests and convince stakeholders of the need for the project
- Make sure that every project has a need (e.g., a problem to be solved or an opportunity to be seized) at its root
- Identify legal and administrative factors that can impact the project as early as possible
- For high-risk projects, engage in more detailed project planning than for those projects of lower risk
- Manage to the plan and amend the plan if circumstances change, so that the plan is always current and relevant
- Take the time to close projects and identify lessons learned
- Conduct closing discussions for projects that have succeeded and those that have failed
- Engage vendors in closing activities and the identification of lessons learned

DISCUSSION QUESTIONS

1. Who are the influencers of your public-sector projects? What areas have you been drawn into that lie outside of your usual area of interest? What strategies have you engaged in for integrating the influencers of your projects?

2. What do project charter documents look like in your organization? Do you have a charter document? What can you do to build a solid process for project initiation?

3. What activities do you get engaged in that do not support the progress of your project? How much of your time is taken up by activities that only indirectly support the goals of the project?

4. What other impacts are created when simple changes are made in your projects (e.g., a delayed deliverable)? How well are those changes integrated into the project plan?

EXERCISES

1. For a project that you are familiar with, create a project charter using the template provided.

2. Identify necessary elements of the project plan for a public-sector project.

Electing A Candidate

Although we may not think of it in project terms, the campaign to nominate and elect a candidate is a project. It has severe time constraints, because the election cannot be postponed if we are not ready. It has budget constraints, which are imposed by the donors and rules and limits on campaign contributions. It is a high-risk endeavor, and it involves large numbers of largely volunteer team members who need to be managed and coordinated. For high offices particularly, the project can stretch out over many years, with clearly defined phases (e.g., exploratory efforts, the primary, the general election, and the assumption of office). Perhaps more than any other project, political campaigns are driven by networks of stakeholders and communications management.

Fundraising and managing campaign funds is a critical project deliverable for electing a candidate. Although campaign finance laws

(continued)

require detailed reporting and limit the types of expenditures, substantial funds need to be raised and accounted for. For each successive election, it seems that the amount of money required for a successful campaign grows astronomically. Funds need to be raised from businesses and industries, private citizens, professional groups, and others.

Today, campaign organizations operate at the local, state, and national levels and require substantial numbers of paid staff and volunteers. In U.S. presidential elections, the campaign organization used for acquiring the nomination is expanded by recruiting experts from the campaign organizations of former adversaries. By the time of the election, the campaign may involve thousands of staff members.

Detailed campaign plans are also necessary for a successful campaign. They include analysis of the image the candidate wants to project, issues to be emphasized, strategies for employing the media in the campaign, identification of voters likely to support the candidate, assessments of voters and areas most likely to be cost effective for the investment of campaign resources (money and volunteer time), staffing plans for the campaign, and events and schedule coordination. Those plans, though detailed, have to be flexible enough to take into account changing circumstances and the strategies employed by the other candidate.

Today, television advertising comprises the single largest expense in any presidential election campaign. If TV advertising is a major project deliverable, subordinate deliverables include creating TV advertisements, analyzing TV markets and demographics, purchasing the ads, and evaluating ad results and modifying the plan.

Other project deliverables could include preparing for debates, conducting media interviews, obtaining newspaper editorial board endorsements, polling voters, creating positive and negative messages, managing campaign staff, and analyzing the campaign's effectiveness and likelihood of success in the election. Political campaigns also depend on the ability of the campaign leaders to motivate huge numbers of volunteers and donors.

Chapter 5

Managing Project Scope

PROJECT SCOPE MANAGEMENT

Nothing kills public-sector projects (and project managers) more often than poor scope definition. Good scope definition is the foundation for project management, and without it, project failure, frustrated team members, and dissatisfied stakeholders are almost guaranteed. In addition, good scope definition significantly enhances the ability to control project changes as they arise and provides a tool for managing the expectations of stakeholders.

In order to create that good scope definition, a list of deliverables has to be created for the project. The tool used in project management to define project scope is the work breakdown structure (WBS). It can keep scope under control and keep the focus on creating the right things and only the right things.

Project scope management is the foundation of the triple-constraint model. Once scope is identified, the budget and schedule flow from it. With a good scope definition, project phases can also be identified that can help keep the project under control. Scope definition is not a simple process done in isolation by the project manager or the project team. It takes the involvement of stakeholders and the application of sophisticated facilitation, consensus building, and conflict management skills. Although nothing can seem more difficult than building agreement on the scope of a complex project, no effort provides a higher dividend.

THE CHALLENGES OF SCOPE MANAGEMENT FOR PUBLIC-SECTOR PROJECTS

If managing project scope identification is a problem for most project managers, imagine the extra complications of it for public-sector projects. In the private sector, identifying the business goals of a project is relatively straightforward. In the public sector, however, goals are more complex and fragmented. In addition, more stakeholders are allowed both by public-sector rules and political exigencies to have a voice in the direction of projects and agency strategic directions.

Managing project scope is a process that requires broad involvement of stakeholders and the development of consensus on the purpose and deliverables of the project. In the private sector, an organizational authority figure can often serve as the final arbiter of the project's scope. In the public sector, few officials are willing to undertake a project on which consensus among a wide array of stakeholders has not been identified. That unwillingness to unilaterally direct outcomes protects the public from unwarranted action by public officials, but it also complicates project management.

Sometimes, project managers at the agency level—in order to take control and get the project started—attempt to develop a definition of the project's deliverables but neglect to share that definition broadly and make sure that other stakeholders have the same understanding. Without clarification of the goals of the project by the project manager, stakeholders are free to invent their own definition of the project.

For example, if a project were undertaken to revise the funding formula for the distribution of funds to county governments for the care of those with mental illnesses, different expectations of the outcome of the project could include:

- For fiscal staff, a means of simplifying the distribution of funds
- For rival programs and budget officers, a means of reducing the funding to the counties
- For all counties, more money
- For small counties, correction of inequity and distribution of more money to them
- For proponents of specialized services, more money for their specific needs

- For system managers, a better systemwide balance between funding and service provision
- For political leaders, support of their constituents
- For legal staff, a more legally defensible distribution of funds

Failure to create consensus on a clear scope allows each of these stakeholders to form their own opinion of the project and express disappointment when the outcome does not meet their specific expectations. Developing that consensus among all stakeholders can take a considerable amount of time and the utilization of a sophisticated skill set, including skills in communications, consensus building, and mediation.

This business of ensuring that all of the project's stakeholders, including the project manager, have a clear idea of the project's deliverables is called scope definition. It is one of the most critical steps in successful project management. In fact, poor scope definition is, by far, the most common cause of project failure. Scope is usually defined as the range or width of something. In the case of projects, scope is the measure of the breadth of the project. It defines what is included in the project. By defining what is included, it automatically defines what is not. The folly of creating projects with wide scope will be addressed later.

To use a construction metaphor, if the project charter is the foundation for a project, the scope is the frame and roof. Everything in the project has to fit inside the boundaries established by the scope. The frame has to fit on the foundation, and everything done after that has to be accomplished within that frame. Project scope identifies the range of things a project is supposed to create. It includes the listing of deliverables (nouns) that define a project's intended outcomes. In truth, nothing is harder than coming to grips with project scope. If that is right, the rest of the project is easy (or, at least, easier).

Even in projects that are largely self-contained within the department or agency, developing a good project scope is tough. For example, assume that a training program in information security awareness is being created for the staff of a large public agency. The overall goal is to reduce the probability of information breaches and the impact of those breaches if they occur. In order to develop the scope for that project, the following questions need to be answered:

- How long is the training? (one day, a week?)
- What training options are available? (online? instructor-led?)

- Will the IT staff provide content, or does it have to be developed?
- Do existing security policies have to be included in the training? Are there existing policies that can inform the training?
- What is the current level of security awareness of employees? What assumptions can be made about their level of awareness?
- Can training be purchased off-the-shelf?
- Do training attendees need to be tracked? Do they need to be graded or tested after the training?
- Has training been conducted in the past that can be used for a model? Were there lessons learned from the prior training?
- Does this training have to be repeated annually?
- Does the training need to be reinforced with periodic newsletters or posters?
- Does the training have to be delivered or just developed?
- Does the training need to be evaluated after it is delivered to determine its effectiveness?
- Are there time constraints?
- Is there a budget?
- Is the same training to be delivered to all staff? Do IT staff members need specialized training?

Each combination of answers will define a very different project. The chore is to figure out what is expected from that range of project possibilities. In this example, the project could be as simple as evaluating and purchasing off-the-shelf training and turning it over to offices for delivery, or as complex as identifying current knowledge, building various curriculum sets that meet different needs and address subordinate departments, integrating delivery mechanisms, delivering the training, testing those who have been trained, tracking attendance, evaluating the impact of the training, and repeating the training annually.

The best project management methodologies understand that project scope must be defined before project activities are identified, and that activities must be directly connected to deliverables to prevent the performance of activities that do not contribute to the project's success. To create the full listing of project deliverables that can assure consensus and serve as a basis for the identification of activities, a work breakdown structure (WBS) must be created, which will become the most important piece of the project plan.

Best practices in project management require that the scope of projects be limited. Broad projects with less-than-clear scope have little chance of success. The chances of project success are increased by creating a series of short, clearly defined projects. For example, a public agency might determine that IT staff exhibit high rates of turnover and that managers have a difficult time recruiting skilled IT staff. The agency might then identify the need to review the current classification system for IT staff, compare the current rates of pay to the market, create a new set of classifications, fit those new classifications into a system of compensation, notify all IT staff of the change, and execute the change. In order to optimize its chances of successfully implementing a change and solving the problems it faces, the agency should break this effort into a series of projects rather than engage in one broad project designed to accomplish all of these activities. As each individual project was completed, the agency would have the opportunity to base the next project on the successes of the prior one and to revise the project requirements as necessary dependent on prior projects.

THE TWO ROLES OF PROJECT SCOPE

Project scope management has two components. The simplest is its definition of the scope of the *project*. The scope of the project defines what deliverables are to be produced by this project. For example, a public-sector construction project may include a feasibility study to determine the costs and benefits of the project. However, a feasibility study might not be conducted or might be conducted before the start of the project as a stand-alone project of its own. One project-based deliverable that should be included in every project is a project plan, which will include the schedule, the budget, and other planning outputs that are described later.

Similarly, a project to improve processes for approving travel requests might contain the evaluation of the new processes after they have been in operation for a while. Conversely, that evaluation might be part of another project or a separate project on its own. To develop the plan for a project, the start and finish must be defined.

The second role of project scope has to do with the features and functions that define the *product* (e.g., whether or not the enterprise accounting system being deployed for the state will contain a module for issuing warrants). Product scope defines what the product or service produced by the project will look like and what features it will contain.

THE REQUIRED FUNCTIONS FOR PUBLIC-SECTOR PROJECT SCOPE MANAGEMENT

Earlier, the four required functions for the management of scope within a public-sector project were identified. They were:

- Creating a scope management plan
- Defining and confirming project scope
- Verifying acceptance of deliverables
- Controlling scope

Each of these functions will be discussed in turn and identified as to how they apply to public-sector projects.

Creating a Scope Management Plan

Creating a scope management plan defines how the project team will manage scope. It creates a scope management plan that can:

- Describe the activities engaged in to manage scope
- Describe processes for approving scope changes
- Identify the stakeholders who will participate in scope management and those who must approve changes
- Identify the processes by which stakeholders accept or reject project deliverables

Given the complexity and risk of the project, the project team needs to decide how detailed the scope management plan will be.

A template for a scope management plan for a public-sector project follows in Table 5.1.

Defining and Confirming Project Scope

Defining and confirming project scope continues the process of progressive elaboration of the project. In some projects, a scope statement is prepared

Table 5.1 Scope Management Plan Template

Project Name	
Project sponsor	Identify the person or organization who has ultimate scope authority
Customers	Identify the persons who have the ultimate authority for acceptance of project deliverables
Scope approval process	Identify the process for presentation of deliverables to customers and the methods by which they will confirm acceptance or indicate deviations from expectations
Process for creation of the WBS	Identify the processes that will be used and the persons involved in the creation of the WBS
Process for the acceptance of the WBS	Identify those people or groups who are involved in the review of the WBS and the process by which the WBS is to be signed off on
Process for submission of proposed scope changes	Identify how project team members and stakeholders can request a scope change, including creation of the forms to be used and the data to be provided
Process for reviewing and approving scope changes	Identify the process by which requested scope changes are reviewed and approved to include participants in that process and methods by which decisions will be made.
Integration of changes into the project plan	Identify who has responsibility for integration of changes into the project plan
Scope change communication	Identify the communications methods and the persons responsible for communicating scope changes to stakeholders
Contract change mechanisms	Identify the methods by which contracts with vendors will be amended to accommodate scope changes

after the project charter but before the development of the WBS. Other organizations simply use the WBS and the charter to complete the definition of the project's scope. The ultimate goal of these documents is to describe the deliverables of the project and describe how the project team intends to create those deliverables. It, like the other project documents, should be circulated to and signed off on by stakeholders.

Additional project details that could be added to the charter or documented within a scope statement are:

- High-level deliverables
- Preliminary project risks
- The organization of the project
- Reporting requirements
- Key stakeholders
- Sign-off requirements
- More detailed listings of constraints and assumptions
- Any other details that have become known through progressive elaboration during the time elapsed since the creation of the charter

Ultimately, the project team needs to create the full definition of the project's scope by creating a WBS. A WBS is nothing more complex than a hierarchical listing of the deliverables for the project. It can be created either in outline form or graphically. It is built through the process of decomposition, which is the successive breaking down of deliverables into smaller and smaller pieces. No matter what form is used or how far the WBS is decomposed, it has to include all of the deliverables of the project and no deliverables that are not part of it. Each of the deliverables needs to be stated in terms that can be understood by all project stakeholders.

For example, assume that an enterprise-level information system is being built for an agency that has been created to regulate and set water prices for a nation. At the highest level of the WBS is the ultimate product of a project (see Figure 5.1). An outline format is initially used for the WBS. A graphical format for the same project is included later.

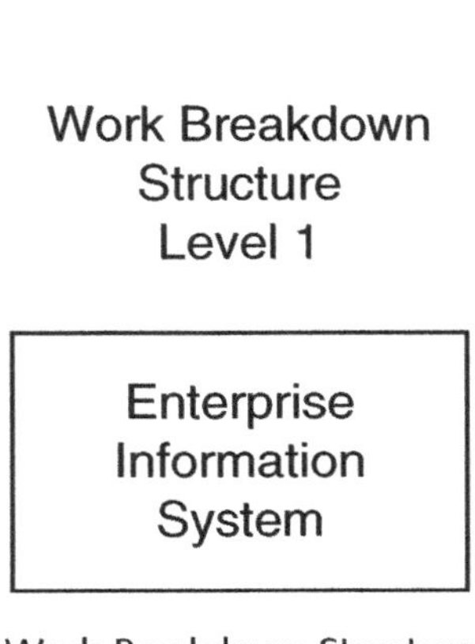

Work Breakdown
Structure
Level 1

Enterprise
Information
System

Figure 5.1 Work Breakdown Structure—Level 1

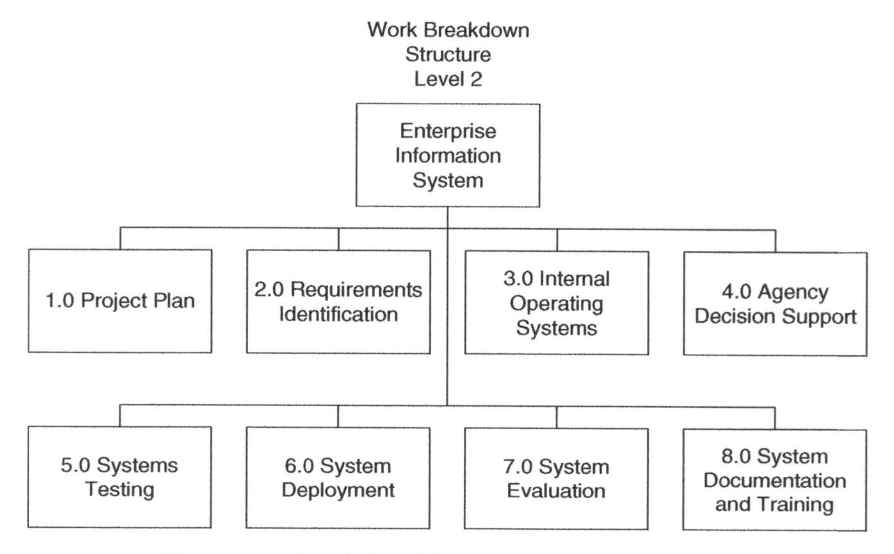

Figure 5.2 Work Breakdown Structure—Level 2

The second level of the WBS becomes much more interesting, because it begins to break down that overall project deliverable to provide more detail. The second level of a WBS for the enterprise information systems is shown in Figure 5.2.

Note that the WBS is a mix of project deliverables that are for the project team's use only (e.g., the project plan) and system components (e.g., internal operating systems) that are of use to stakeholders. Deliverables in the WBS do not need to be put into any logical sequence; project activities will be sequenced later.

With the second level of deliverables identified, the project is scanned for logical phases to help manage the project and communicate about it. For example, it may make sense for this project to create a "planning and requirements identification phase," which would combine those two deliverables into one manageable phase. Note that deliverables are described using nouns and are numbered so that they can be linked to the higher-level deliverables that spawned them.

From here, each of the deliverables is further decomposed. The third level of the WBS illustrates that further decomposition (only a couple of the level-two deliverables are shown), and an outline format instead of the graphical format is used for the first two levels (see Table 5.2). The choice of the format is optional and should depend on determination of the best way to communicate with stakeholders.

Table 5.2 System 5.0 Testing

5.1	Test plan
5.2	Test requirements
5.3	Test environment
5.4	Testers
5.5	Test implementation
5.6	Test evaluation

The other deliverables are further decomposed by continuing to move to lower and lower levels of the WBS. For example, the deliverable 5.4, Testers, could be decomposed as in Table 5.3.

The WBS has several purposes. First and foremost, the WBS helps determine what needs to be produced and keeps the focus on deliverables so that random activities that will eat up project resources and time without contributing to the creation of necessary deliverables are not performed. It creates a mental discipline to help explore the project in an orderly manner.

The WBS and its associated WBS dictionary, which defines terms used in the WBS so that everyone can understand what the WBS contains, is the definition of a project's scope. It is the scope baseline, which is the original scope plus *approved* changes. Scope changes are addressed later.

Second, the WBS is used to organize the approach to the project by identifying phases. At the end of a phase, project deliverables will have been created that can be evaluated by project stakeholders to determine whether or not they are acceptable. Phases also provide the opportunity to review the project and its intended objectives to determine whether to proceed to the next phase. In some cases, at the end of a phase, the project may be terminated entirely if the business need for which it was initiated is no longer valid.

Table 5.3 Deliverable 5.4 Testers

5.4.1	Identification of testers
5.4.2	Scheduling of testers
5.4.3	Training of testers
5.4.4	Debriefing of testers
5.4.5	Follow-up with testers

Third, the WBS provides a tool for communicating with project stakeholders. For example, stakeholders can be briefed about the WBS. As deliverables are identified for them, they might suggest changes, which is a good thing in most cases. Projects can always benefit from the input of others, and with that input, the WBS may be modified. After decomposing, discussing, and revising are completed, a solid, agreed-to WBS will be in place. Once that WBS is in place, a much better base will exist from which to manage discussions with stakeholders later in the project about the changes they might want to make. (If the project's scope is not defined in detail, changes in the scope that will be requested later cannot be managed.)

The WBS is fully decomposed when the deliverables at the lowest level of the breakdown are understandable and can be converted into the activities that will be required to create those deliverables. Deliverables at the lowest level of any WBS branch are called work packages. Note that deliverables do not have to be broken down to the same degree on all branches of the WBS.

Controlling Scope

Scope control is a subset of monitoring and controlling the project, which was described earlier. It is addressed here because of the importance of scope control as a means of ensuring that the objectives of the project are met. By limiting the project's scope, chances of success are increased. Prior experience has shown that the wider the scope, the higher the probability of failure. Creating a clear project scope presents another advantage later in the project, because it can be used to help manage the project and make better decisions about it down the road.

Over the course of the project, stakeholders may impose pressure to expand the scope. For example, in the information system project, some stakeholders may exert pressure to include other system modules and add other system functions. During information system design, it may become clear that the processes it will automate are poorly developed and require modification. Evaluating and creating new processes may be a logical addition to the project. If project scope is clearly defined, the impact of changes to it can be examined and informed decisions made about whether the proposed change can or should be accommodated.

The uncontrolled addition of scope to a project is called "scope creep." Avoiding scope creep does not, however, imply that scope should never change. Some scope changes should occur as the needs for the project and understanding of those needs change. It is unlikely that all of the project's deliverables could be known at the start of the project, if for no other reasons than changing circumstances that become known as the project progresses. However, logical decision making should guide scope changes, and the impact of those changes should be clearly known when those decisions are made.

Without good scope control, stakeholders may try to increase the scope of projects to include work that they would like to have done or work that they just discovered they need to have done. In some cases, those stakeholders may be trying to reverse scope decisions made earlier that did not deem their needs to be a high priority.

For example, in the revision of the agency's accounting systems, the highest priorities may have been assigned to those deliverables that contribute to fulfilling reporting requirements to higher-level agencies. Some agency managers, however, may have been interested in creating a system to provide better data for decision making. Although the needs of those managers may not have been assigned a high priority in the construction of the new system, they may try to change the scope definition process later to include their needs in the project.

In order to address stakeholder demands, a clear definition of the project scope and a solid scope change management process are needed. Although it might be tempting for the project team to exercise decision authority for scope changes, the team will be better protected if the stakeholders are involved in scope change decisions.

Scope change processes are also employed when scope needs to be reduced, because of a lack of resources or time or because other scope has been added to the project. Reducing the scope of the project is not something that the project team should attempt on its own. Instead, the project team should employ broad stakeholder input and a set of decision rules and processes to determine how scope change requests will be evaluated and who has the authority to make scope changes. The ultimate outputs of scope changes are revisions to the project plan, including changes in the WBS, which, as noted, is the definition of the project's scope.

Verifying Acceptance of Deliverables

The achievement of the project's scope is ultimately verified by turning over the product or service created by the project to its stakeholders and determining if they find those results acceptable. Note that there is a difference between *acceptable* results and *correct* results. Correctness of results will be discussed along with project quality. Correctness is a higher standard than acceptability.

Getting stakeholders to sign off on results can be a challenge. Some organizations have a culture in which sign-off is an accepted part of the transfer of deliverables. In public-sector projects, getting stakeholders to sign off on project results may be a challenge, particularly given the wide array of stakeholders in public-sector projects and the time it may take to conduct a full review of results. It is not improbable in a public-sector project that a key stakeholder or stakeholder group will notify the agency or project team that errors have been found months after the creation of a deliverable. That is particularly the case with public-sector oversight agencies, which may conduct periodic structured reviews and audits of agency processes.

Even without sign-off, stakeholder acceptance of deliverables can be confirmed by e-mail or by making a notation in the project record of acceptance.

BEST PRACTICES FOR MANAGING PUBLIC-SECTOR PROJECT SCOPE

Best practices for managing the scope of public-sector projects include:

- Do not move forward without a solid project charter
- Make sure that everyone understands the problem that the project is intended to solve or the opportunity that is being seized
- Use project assumptions to protect yourself and the team (e.g., conspicuously note the responsibilities of other offices as project assumptions)
- If a project charter is not provided, build one and circulate it to stakeholders
- Include necessary permissions and approvals in the scope of the project
- Build in time and resources for the review of rules and policies
- Build a WBS and require stakeholders to understand it and agree to it
- Make sure that all stakeholders understand the scope change management process
- Do not take responsibility for making decisions about scope changes; instead, rely on the process

- Make sure that project team members do not make scope changes on their own without authorization
- Get sign-off on deliverables
- If the organization does not have a culture that supports sign-off, notify stakeholders that they have a limited time to review the product before their sign-off is implied
- Remember that time and cost management depend on scope management

DISCUSSION QUESTIONS

1. What challenges do you face in managing project scope for public-sector projects?
2. Do your projects create a scope management plan? What elements would be useful to include in a scope management plan if you were to create one? Do you have standard organizational processes for scope management?
3. What processes do your projects use for determining whether or not to accept proposed scope changes? Where do those changes come from? What could be done in your organization to better manage scope changes?
4. How do you hand off deliverables to your stakeholders? Do they sign off on acceptance? How could you improve that process?
5. What other best practices can you imagine for managing project scope in public-sector organizations?

EXERCISES

1. For a project that you can imagine embarking on, identify the questions that you could ask to help define project scope.
2. For that project, build a scope management plan.
3. For a public-sector project you are familiar with, create a WBS, decomposing it as far as necessary.

Projects for Improving Public-Sector Processes

Public agencies often depend on processes for performing work. Those processes are used to issue licenses and permits, process tax returns, determine the eligibility of applicants for public services, issue payments, process inspections and audits, and perform a host of other functions. Improving those processes demands the constant attention of public agencies.

In order to improve processes, public agencies need to initiate projects. The most promising tool today for improving public-sector processes is Lean government. Lean government combines statistical process control methods, the structured elimination of waste, value-stream mapping, and Kaizen events, which are structured and intense two- to five-day exercises designed to focus on a specific process. Kaizen events bring together diverse stakeholders using a skilled facilitator with the goal of accomplishing as much as possible in a short period of time.

As an example, Lean government projects were initiated beginning in 2003 in five state agencies responsible for environmental protection. The results they achieved were substantial. Some examples of results were:[1]

- Lead times for permitting processes were reduced by more than 50% without reducing time for substantive review.
- Permit backlogs were reduced or eliminated; in Michigan, permit processing time for major air construction permits was reduced from 422 to 98 days; in Iowa, the average time necessary for issuing a standard air-quality construction permit was reduced from 62 to six days; in Delaware, the backlog of air construction permits was reduced from 199 to 25 days.
- Processing time variability was reduced and consistency was improved.
- More staff time was made available for other critical work.
- Customer satisfaction was improved.
- Staff morale was improved.

Critical elements in the management of Lean government projects are the establishment of the boundaries of the project (i.e., what

(*continued*)

processes were out of bounds and what critical functionality needed to be protected in the public interest) and the creation of solid metrics to support the process improvement process. The project manager for a Lean government project is also challenged to create excitement about the change process, while 1 ensuring that the team remains realistic about what can be accomplished.

[1] United States Environmental Protection Agency, "Working Smart for Environmental Protection: Improving State Agency Processes with Lean and Six Sigma," September 2006.

Chapter 6

Managing Project Time

THE CHALLENGES OF PROJECT TIME MANAGEMENT IN THE PUBLIC SECTOR

Private-sector projects often have time constraints. A company may want to introduce a new product as quickly as possible, or a process may need to be improved on a the-sooner-the-better basis. Private-sector projects may also need to comply with external deadlines for compliance with laws or government regulations. Public-sector projects have some of the same time constraints. However, public-sector projects also are often affected by more rigid time constraints imposed by:

- Statutory requirements for program execution
- The requirements of due-process protection and speedy resolution of claims, complaints, and filed comments
- Statutory limits on processing times
- The rush to accomplish an agenda during the relatively short term of elected officials
- Budget and spending time limits (i.e., money for projects can expire at the close of a fiscal year)
- The requirements of higher-level agencies (i.e., the U.S. federal government often sets time constraints for project performance for state agencies)

Although some people assume that the public sector moves slowly, the time horizons for planning and for execution are often shorter than they are in the private sector. For these reasons, public-sector agencies and managers have very low tolerance for schedule risks and project delays, and public-sector project managers are often constrained by what can seem like arbitrary due dates for project deliverables.

THE REQUIRED FUNCTIONS FOR PUBLIC-SECTOR PROJECT TIME MANAGEMENT

Two required functions were identified for project time management for public-sector projects. They were:

- Defining project activities
- Creating the project schedule

Each of these functions is discussed in turn.

Defining Project Activities

Defining project activities is a follow-on function to the creation of the WBS. The WBS was composed of the nouns/deliverables of the project. At the lowest level of any branch of the WBS are work packages, which are still defined as nouns/deliverables. Some WBS samples found in other publications contain activities as well as the deliverables.

We know that we have decomposed the WBS as far as we can when we are ready to start identifying the activities/verbs that need to be undertaken to create those deliverables. For example, in the example used in the last chapter, one of the deliverables was system hardware. If we conclude that we do not want to decompose the deliverable any further, we can create a list of the activities necessary to create the deliverables. For the hardware deliverable, we may want to decompose it into activities as shown in Table 6.1.

The outcome of activity definition is an activity list. It could help us to understand the process of making the conversion from nouns to verbs by imagining that we created the WBS on a whiteboard. When we have finished the WBS on that whiteboard, we could start creating activities on Post-it® Notes. Although we would place those Post-it® Notes on the whiteboard with a direct connection to a deliverable, we could pull the Post-it® Notes off of the board and stack them up. That stack of Post-it®

Table 6.1 Decomposition of Activities from a Work Package

3.5 Hardware	
3.5.1	Analyze hardware needs
3.5.2	Identify hardware options
3.5.3	Create a recommendation for purchasing hardware
3.5.4	Contact purchasing department to initiate purchases
3.5.5	Identify possible vendors
3.5.6	Select vendor
3.5.7	Order hardware
3.5.8	Inspect hardware on delivery

Notes is our activity list. Now that we have that list, we can start to manipulate the activities using the processes that will follow. Ultimately, we will assign attributes to those activities, which can include the number, description, predecessors and successors, resource requirements, constraints, and other factors.

It is important to be able to identify each activity with a specific deliverable, which is done by creating a numbering system for deliverables that is continued through the activities (the code of accounts). Any activity that cannot be associated with a deliverable is not a part of the project. When an activity is not directly connected to a deliverable, we have the choice of eliminating it or amending the WBS. Modification of the WBS, however, is a scope change in the project, because the WBS defines the scope. To modify the WBS, we may have to invoke scope change management processes, which were identified in the scope management plan, the principle output of scope management planning.

We can tell if we have broken the activities down into fine-enough detail if they can be budgeted and scheduled and are complete and necessary. Some project managers apply the 8/80 rule for activity definition, which states that activities should take no more than 80 hours to complete or no less than eight. In reality, the degree to which activities are broken down is dependent on the project and the capabilities of the project team. If a project team member is inexperienced and untested, that team member can be assigned to shorter activities so the project manager can get feedback on their performance faster.

We can also engage in rolling-wave planning, which creates more project detail for the activities that will occur in the short term, while leaving those activities that will occur in the long term at higher levels of generalization for the time being. Rolling-wave planning derives directly from the concept of progressive elaboration, which states that we will learn more about the project as we go.

You might be aware that most project management software begins with identifying project activities or tasks. To some extent, they lure us into the project at the wrong point. Good project management would demand that we identify deliverables *before* we attempt to enter activities or tasks. As long as we understand that activity definition is not the starting point for project planning, the software can provide us with valuable assistance in tracking project details.

For the assignment of activities to staff, we need to make sure that the activities are clear, measurable, and appropriately assigned given the expertise and experience of the person or persons to whom they have been assigned.

In public-sector projects, we need to take special care to identify the activities related to:

- Interfacing with purchasing and human resource processes and offices
- Interacting with the wide array of project stakeholders, particularly those who can influence the outcome of the project
- Researching legal and administrative impediments and processes that bear on the project
- Crafting amended rules, processes, and legislation
- Participating in budget and project reviews

Creating the Project Schedule

Now we have a stack (or list) of project activities. So what? We have not created a schedule and cannot do so until we have done a few more things. We have to identify the correct order in which those activities should be performed, determine how long they will each take, and pull them together to create a workable project schedule. In addition, we will also need to identify the resources needed for each of those activities, just to be certain at this point that there are no resource conflicts that could impact our schedule. Those activities can performed in any order that makes sense, but they have to be integrated. Each of those sets of activities is described as follows.

Putting the Activities in a Logical Order

In order to determine the right sequence of activities, we have to ask ourselves a simple question for each one: What activities have to be done before this one? That is called identifying dependencies. It tells us which activities are dependent on other activities.

Some dependencies are mandatory. That means that we *have* to do some activities in a required order, like installing the new system for issuing permits only after that system has been developed or purchased. The laws of the universe and logic prevent us from doing those activities in the reverse order.

Other activity relationships are discretionary. That means that we *prefer* to do them in a certain order. Without thinking about it, those preferences for doing things in a certain order are usually related to reducing a risk that we are not comfortable with. In some cases, discretionary dependencies represent industry best practices.

For example, a state health department annually submits proposals for bioterrorism grant funds to the Centers for Disease Control and Prevention (CDC), which evaluates them for federal government funding. The typical sequence of activities for submitting those grants is:

- Identification of grant ideas
- Creation of a short business case for each grant idea
- Submission of the short forms to the management of the department and the governor's office for approval
- Review of grant short forms by the department's managers and the governor's office
- Preparation of full applications for those grant proposals approved by the department's management and the governor's office
- Submission of the full grant proposals to the CDC for its evaluation and selection for funding

That might be a logical and preferred sequence, but the CDC has a firm deadline date. If the state health department misses that deadline by an hour, the grant proposal is automatically rejected.

In that grant submission process, there are both mandatory and discretionary dependencies:

- Mandatory: The grant proposal cannot be submitted before it is written.

- Mandatory: According to the rules of the state, the proposals have to be reviewed by the management of the department and the governor's office before they are submitted.
- Discretionary: The staff of the department will not write the full proposal documents until the short descriptions have been approved by the department's management and the governor's office.

Given the high level of schedule risk (the firm deadline date for submission to the CDC), the department may want to reconsider the discretionary dependency in order to reduce that risk. To save time, they can start writing the full proposals while the short proposals are being reviewed.

If we relax the discretionary dependency and adopt this option, we are making a risk trade-off. Our original discretionary dependency (waiting for review before investing the time to create the full proposal) was chosen in order to reduce the risk that we might waste time by writing a full proposal that might not have passed the management and governor's review. We can, in this case, choose to reduce the schedule risk and increase the resource risk.

We can further classify activities as predecessor activities or successor activities. Predecessor activities have to be done before the next one. Putting on our socks is a predecessor activity of putting on our shoes. Successor activities say the same thing in reverse. Putting on our shoes is a successor of putting on our socks. An activity can have more than one predecessor or successor.

Once we have analyzed the dependencies among tasks, we can put them into a project network diagram. The easiest way to create the project network diagram is to put the activities on pieces of paper that can be shifted around as we think our way through the diagram. As noted earlier, some people like to use Post-it® Notes. Alternatively, project management software can help us organize activities and sequence them in complex projects. There is no single, right way to sequence activities for any project. What seems most logical to us might not seem logical to other people, and we may even change our minds about the right sequence later. It is a matter of doing our best with the information we have.

There are two methods for mapping project sequences and creating a project network diagram: the activity-on-node model (AON) and the activity-on-arrow method (AOA). The AON model places activities on the nodes of the diagram. The AOA model puts the activities on the

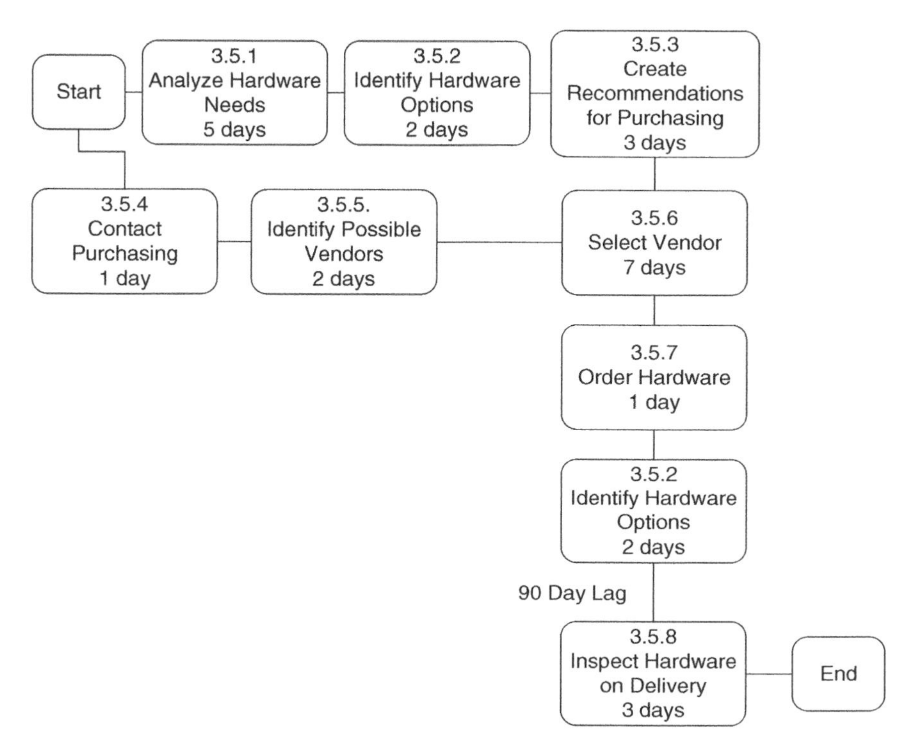

Figure 6.1 Project Network Diagram

arrows. Most organizations prefer the AON model, which is also referred to as the precedence diagramming method (PDM).

A project network diagram using the AON model is illustrated in Figure 6.1.

With the constructed network diagram, we can literally *see* the link between activities and are very close to being able to build our schedule. On this diagram, duration estimates have been added, which are described in the next section.

Note that a lag has also been added between activities 3.5.7 and 3.5.8. That lag indicates that, even though 3.5.8 follows 3.5.7, it has to wait an additional 90 days after 3.5.7 has been finished before it can start. (We have to wait for hardware to be delivered by the vendor from whom we purchased it before we inspect it.) A lag like this can be used to create a network diagram that is more reflective of realistic time constraints.

A lead between two activities is the reverse of a lag. It indicates that a successor task can begin before the predecessor has been fully completed. For example, the votes cast in an election include ballots cast at the polling place and absentee ballots. In most cases, although absentee ballots have to be postmarked by the day of the election, they may not be received for up to two weeks after that. Even though the final election results cannot be confirmed until the absentee ballots have been received, the vote counting can begin as soon as the polls close on Election Day.

Estimating How Long Each Activity Will Take to Complete

If we embark on activity sequencing after activity definition, we will not have determined yet the duration of activities—how long each one will take. We may have identified the time constraints of the project earlier, but that does not mean that the activities identified as being necessary for the project can be completed within those time constraints.

Activity duration estimating identifies the time necessary for completion of the activities identified. We are not concerned with how long an activity *should* take without the other demands on the time of our resources; we want to estimate how long it *will* take them after considering their other assignments and duties. For example, a key resource may only be available to the project half-time (20 hours per week). If that person were assigned an activity that would normally take 40 hours to perform, we would assign a duration of two weeks to the activity rather than one week.

After we estimate durations and match those durations to the project network diagram, we can determine the total length of the project. That project duration can be compared to the project's time constraints to see if we have a realistic plan. We may need to reconsider our project's scope or consider compressing the schedule to make it fit within the time allowed for the project.

There are several methods for estimating the duration of activities. These methods vary based on the amount of information that is required for the estimate and their level of accuracy. The least accurate estimating method, which requires very little project data, is analogous estimating. In that method, we simply find another project or activity and identify the length of it. We can then make simple adjustments based on the differences between that project and the current project and derive an estimate. Analogous estimating is often used to provide early, rough-order-of-magnitude estimates to establish budgeting parameters.

If we have a little more project data, we can identify mathematical relationships to help estimate durations. This is called parametric or quantitatively based durations. Parametric estimates are regarded as more accurate than analogous estimates because they require more data. For example, if we were to attempt to determine how long it will take us to write a software program estimated to require 100,000 lines of code, we could find out what the standard is for the number of lines of code that can be written in an hour and divide the total lines necessary by the standard to determine the number of hours the project will take.

We can also gather independent information or rely on experts to help create estimates. If we rely on expert judgment, we can minimize the impact of very high or very low estimates by using a three-point estimate. A three-point estimate identifies the most optimistic estimate, the most pessimistic one, and a midrange one. Those three estimates are added with the midrange estimate being multiplied by a factor of four. The sum is then divided by six, which pulls the high and low estimates toward the midrange estimate.

The most accurate estimate may be the one that relies on the most project data, which is called bottom-up estimating and assigns durations to individual activities. Those individual activity estimates, although they employ the most project information, can be inaccurate, however, if the same bias is applied to each activity estimate.

The best method for creating good estimates of duration for a project is probably the combination of several methods. When activity-based, bottom-up estimates are made and summed to create a total project budget, that sum can be evaluated using analogous estimates, which provide a high-level check on the accuracy of the summed estimates. Ultimately, we want to make sure that durations are assigned to individual activities so that we can appropriately track project schedule performance.

In public-sector projects, some of the factors taken into account in assigning project durations are:

- The amount of time required for administrative processes
- The timing of budget cycles
- The timing required for changing rules or statutes
- The timing of review and approval cycles (e.g., if we need approval for a purchase and that approval is subject to review by a standing committee, we need to build the timing of the meeting of that committee into the project's duration estimates)

Identifying the Necessary Resources for Each Activity

The last piece we need to put in place before we can create the project schedule is activity resource estimates. Those estimates define the resources that are necessary for completing each activity. It may seem that identifying resources is more appropriate to project budgeting and cost management. In fact, in identifying the necessary resources, we are not concerned with the cost of those resources. We are only concerned with identifying what resources are necessary for the performance of the activities of the project.

Identifying necessary resources is critical to the development of the schedule, because the timing of the availability of the resources we need may impact the schedule. For example, we might need an expert in construction of a website, but the only available resource might be assigned to other projects for the next six weeks. In addition, the resources may not be available without our undertaking additional activities. For example, if we were to determine that an expert necessary for the project is not on the payroll of the agency, we would have to include additional activities for recruiting an expert, hiring a consultant, or outsourcing one or more deliverables. In public-sector projects, those additional activities can add significantly to the duration of the project.

In some cases, we may need to reschedule activities to make them coincide with the availability of resources. That problem has given rise to the theory of constraints, which makes the assumption that project planning is done under circumstances of constrained resources. Because of those constrained resources, we have to build the schedule around resource availability, rather than around an idealized schedule based on activity dependencies.

One tool for the management of resource nonavailability in scheduling is the use of buffers in the schedule. Schedule buffers are additional blocks of time placed in the project network diagram before activities or sequences of activities that are resource constrained. Essentially, buffers, which are added to the project duration, indicate that the activity must wait a specified number of days until the resource becomes available. Buffers can also be used to identify the summed differences between an aggressive and a likely duration for a group of tasks if scheduling is based on the theory of constraints.

Constrained resources are one of the reasons for the project adage that "bad news moves forward but good news doesn't." That adage implies that, if an activity is delayed (bad news), the entire project may be delayed. But if an activity finishes early, the next activity still cannot start, because it is waiting for resources. For that and other reasons, we should apply a conservative bias in project scheduling. Optimism is probably going to just get us into trouble.

There are few tools for activity resource estimating. In general, the appropriate team members meet and attempt to identify the resources needed for each activity. We could do resource estimating at a higher level of generalization than the activity level, but if we did, we would not be able to identify the schedule impacts of resource nonavailability. We can use expert judgment or any other source of information that could help us determine what resources we might need for an activity.

The principal outcomes of activity resource estimating are the resource requirements for each activity and, perhaps, a resource breakdown structure (RBS), which is a hierarchical summary of resources necessary for the project.

For a public-sector information technology project, the RBS might look like Figure 6.2.

After identification of the resources, we may need to engage in resource leveling. Resource leveling looks at the assignment of resources

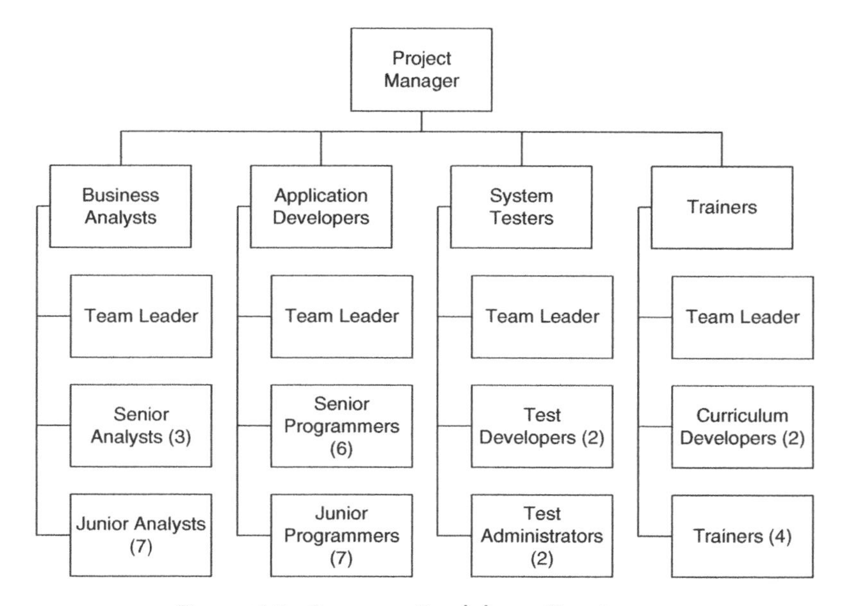

Figure 6.2 Resource Breakdown Structure

across the project or a set of projects to make sure that we have not overcommitted any resources, including people. In the public sector, where project assignments are not tracked well, we often overcommit our most valuable staff by loading work on them without regard to what level of work they have already been committed to. The result is late work, burnout, and the loss of the best people.

In public-sector projects, best practices for activity resource estimating would include:

- Being realistic about the time required for accessing resources from outside the organization
- Being realistic about the availability of resources at the times necessary, especially if those resources are also working on other projects
- Being aware that changing priorities may make previously scheduled resources unavailable
- Documenting assumptions about resource availability in the project scope statement and charter and making those assumptions known in project status reports

An additional challenge in public-sector projects is ensuring that the duties required of the staff person are included within the job description for that individual. Although private-sector organizations have more flexibility in the assignment of duties to staff members, public-sector organizations are bound by civil-service requirements and job rules that require a match between the work assigned and job descriptions and the position assigned to the staff member. For example, a person classified as an Administrative Assistant probably could not be assigned to manage a major project phase. If the project manager wants that person to perform that role, the person will need to be reclassified, which could take significant time and might not be successful. Those rules were intended to make certain that employees are appropriately classified and paid. Although they are well-intended, they create an additional constraint for project managers.

Creating the Schedule by Pulling the Pieces Together

Now that we have identified the activities necessary for creating the project's deliverables, put them into a logical sequence, assigned durations to them, and explored the schedule implications created by identifying resource needs, we can look at the calendar and create the project schedule. We need to examine the schedule in order to identify which days we intend to work, when holidays will occur, and any other days on which work will not occur. There may also be imposed dates that fall within and at the end of the project that could impact the schedule. For example, any project impacting election systems or processes must be completed, tested, and deployed by Election Day.

As was noted earlier, the project schedule is initially prepared without regard for schedule constraints. Only after the length of the project has been computed based on activity durations is it matched to the constraints imposed on it. The most common method of identifying the length of the project is the critical path method, which identifies the path through the project network diagram that is composed of the activities with the longest total of summed durations. That longest path is the critical path, so named because every activity on the path, if delayed, delays the entire project.

Our earlier project network is shown in Figure 6.3 with the critical path highlighted.

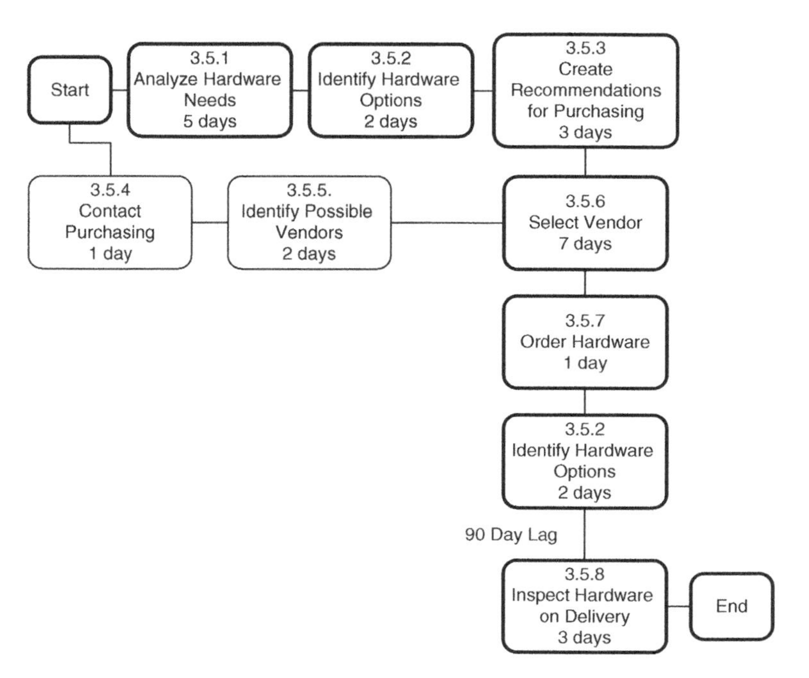

Figure 6.3 Project Network Diagram with Critical Path

If any of the deliverables on that path are delayed, the entire project will be delayed. Conversely, any activity not on the critical path can be delayed without delaying the project. The amount of time that an activity can be delayed without delaying the project is called slack or float. The amount of slack or float for each activity can be calculated. For any activity on the critical path, the slack or float is zero. That is just another way of saying that a delay in any activity on the critical path will delay the entire project. In our diagram, only two activities have slack.

One of the biggest challenges in completing projects is balancing several projects at once. Most people have multiple projects in progress simultaneously. As if one project were not hard enough, balancing multiple projects can make you want to pull your hair out. The good news is that, with a little planning, we can do much better than we might expect.

In the public sector, it is particularly important to keep in mind that the schedule is just a model or best estimate of how the project will unfold. Complex projects defy accurate prediction, as discussed later. Good planning needs to be coupled with project flexibility so adjustments can be made to accommodate changing circumstances and the inevitable delays in projects. Though unnecessary changes to the project schedule need to be avoided, openness to the need for changes is essential, and all of the goals of the project and the risk preferences of stakeholders must be kept in mind. The need to change the project schedule is usually not an indication of bad planning but an indication of changing circumstances and needs.

Compressing the Project Duration: What Happens if the Schedule Will Not Fit within the Time Available?

It is great to have a solid project schedule that gets our project done in eight weeks—unless the due date is six weeks from now. If that is the case, we have four strategies we can use to make the project fit the time available for it. We will look at each of them.

First, if we can demonstrate that we have done good project planning and have concluded that we cannot get the project done on time, we can go back to the person who set the time constraint and ask them to change it. Begging is never an attractive option, but demonstrating solid logic may help get a reprieve. Asking to move the deadline back is an option we cannot use every time, but if we can back up our claim with good evidence and planning, we might get the deadline extended. We will have to be careful to avoid being labeled as a complainer or as someone who always tries to get more time.

The second strategy we can try is to reduce the scope of our project. We can look at those activities on the critical path and examine the deliverables they are linked to. Maybe we can remove some of those deliverables from the project. Third, we can try to throw some more resources at key activities to reduce their duration. In the world of project management, that is called crashing the schedule.

To identify the best options for crashing the schedule, we have to perform a cost-benefit analysis of the options we have. Our goal is to spend the additional resources in a way that maximizes the impact on the critical path. We are, of course, adding additional cost to the project when we crash the schedule, unless we can find other activities to remove resources from without impacting the project.

Our last option is to fast-track the schedule. To fast track the schedule, we need to evaluate the project network diagram and identify activities that are currently scheduled to be performed in sequence but that could be performed at the same time. Those activities have to involve discretionary dependencies rather than mandatory dependencies. Recall, for example, the Department of Health grant application process described earlier. When the schedule is fast-tracked, another risk is increased, such as the risk that quality will suffer.

We need to keep in mind that there can be more than one critical path through a network. That is, there may be more than one path with the same, longest duration. The existence of multiple critical paths increases the overall schedule risk of the project because more activities, if delayed, can delay the entire project. As we compress the critical path, we may discover that the critical path shifts from the original to another.

BEST PRACTICES IN PUBLIC-SECTOR PROJECT TIME MANAGEMENT

Best practices for time management in public-sector projects include the following:

- Build a plan for managing project time (i.e., how will the schedule be tracked, what milestones exist, who will be involved in schedule decisions, etc.)

- Ensure that activity durations measure the time that will be required to complete the activity, not the level of effort
- Build in adequate time for processes that involve other departments or agencies (e.g., hiring, purchasing, or legal review)
- Evaluate options for duration compression even if the schedule does not need to be compressed at the current time
- Build in buffers for resource availability if critical resources are assigned only when they are needed and have other responsibilities as well
- Make sure that job rules allow a team to perform the duties assigned to them
- Recognize the fact that project delays often occur and that tasks completed early will not likely balance out tasks completed late
- Avoid the creation of a single, linear path through the project
- Recognize that the schedule is prepared with the best information available at the time and will always change

DISCUSSION QUESTIONS

1. What challenges do you face in managing project time in public-sector projects?
2. What other best practices can you imagine for project time management in public-sector organizations?

EXERCISES

1. For a project you are familiar with, decompose several deliverables and identify the necessary activities for creating those deliverables.
2. Analyze the following sets of activities and identify which are mandatory and which are discretionary. What risks are being avoided by the discretionary dependencies, and what risks are being increased?
   - Putting the frame of the house on the foundation after the foundation has been poured
   - Building the frame after pouring the foundation

- Testing individual software components after each one is completed (rather than waiting until they are all done)
- Meeting with oversight agencies before finalizing the WBS

3. Using the activities identified earlier, create a project network diagram for your project using the AON method.
4. For the project, create a summary RBS listing the resources necessary for the project.
5. Calculate the critical path through the project network diagram for the project.
6. For the critical path identified, compress the duration of that path by 10 percent.

The FBI's VCF Project

Not every project is a success. In 2000, the U.S. Federal Bureau of Investigation (FBI) recognized the need to replace its Automated Case Support (ACS) system, which was supposed to allow agents to manage documents related to investigations. It had been developed in-house, utilized many separate applications, and was built on software tools that were regarded as obsolete even when the system was deployed.

The new system was to be called the Virtual Case File (VCF) system. After the attacks of September 2001, the scope of the VCF project was enlarged, which nearly always increases the risk of project failure. The VCF system was designed to help the FBI manage cases and was the third part of the FBI's information technology program known as Trilogy. The first two phases of Trilogy created a secure network and replaced outdated hardware.

The original target date for delivery was December 2003. In November 2003, the Director of the FBI received a demonstration of the system, but the FBI later identified 400 software problems and outlined the necessary corrections to the contractor in April 2004. The contractor agreed to remedy the deficiencies, but at a cost of an additional $56 million and an additional year. Neither cost was acceptable to the FBI. In February 2005, the Director of the FBI

(*continued*)

testified to Congress that the system was not yet available and described the FBI's new strategy in the aftermath of the VCF system failure.

The Director stated that the FBI had been handicapped by several obstacles. He argued that the FBI:

- Did not have a complete set of requirements when the contract was signed
- Was handicapped by a contract based on hours worked and was difficult to manage
- Lacked skill sets in critical areas
- Underestimated the complexity of interactions with the legacy system

As we have noted throughout this book, these are not rare problems in public-sector projects, and when they are combined, they can create spectacular failures that capture the public imagination and contribute to public biases and the belief that government is largely incompetent.

Chapter 7

Managing Project Cost

THE CHALLENGES OF PUBLIC-SECTOR COST MANAGEMENT

There are both structural and behavioral barriers to managing project costs in the public sector. Structurally, public-sector projects face constraints embedded in law and administrative rule. Those cost constraints can include limits on salaries and the types of items that can be purchased, requirements for competitive bidding or minority set-asides, and requirements that expenditures be authorized in a budget duly executed by the appropriate authority.

Unfortunately, other than at the budgetary or appropriations level, cost management is often overlooked in public-sector project management. Too often, public-sector projects operate on the principle that "employees are free," given the argument that they would get paid anyway. Too often, project time-keeping systems are not employed or are only applied to contractors. The result is that public-sector project managers cannot tell what a project actually costs. That allows for bad project investment decisions and does not allow for good evaluation of the performance of the project manager or the team.

Even if project costs are tracked, public-sector project managers often argue that too many costs are not controllable by the project team. For example, relatively high rates of overhead costs may be charged to projects, eating up the project budget with costs that cannot be controlled and that do not have an observable effect on the project. In addition, in public-sector projects, resources may be assigned without the concurrence of the project manager. If those resources are not productive, they can eat the budget without producing results, while the team "carries" excess staff. The result, too often, is burnout of competent staff as other project team members get a "free ride." Management of these disparities is discussed more in the chapter on human resource management.

The public sector is also averse to the use of cost data or the type of financial tools used in the private sector. Some agency managers would argue that "the numbers work in business, but we're different." In truth, it is often more difficult to quantify results in the public sector than in the private sector. It is harder, for example, to quantify the benefits of long-term outcomes of school programs than it is to measure increased sales related to a private-sector project. We could measure, for example, if more students have graduated or if test scores are higher, but we have trouble measuring whether graduates are more employable or do better on the job once employed. It is even harder to measure the effect of education programs on the ability of graduates to perform their duties as citizens. Even if the data are available, it is hard to identify which data we want to measure. For example, should we measure the quality of education programs by the percentages of graduating students who avoid incarceration as criminals or the number who vote and hold a steady job? Though measurement of outcomes is hard in some cases, that does not mean that the tools of financial and cost management cannot be used to improve management of public-sector projects.

In addition to the challenges of public-sector cost management listed previously, there may be some reluctance to find out how much public-sector projects actually cost. In a hypercritical environment in which political opponents may try to seize on any opportunity to criticize agency operations, some managers might prefer to operate below the radar of scrutiny and prefer to answer questions with "we don't know" rather than providing data, especially if they fear that the costs have gotten out of control or might not be politically acceptable. There is also often some concern that creating project cost data will require sharing salary information with a wider subset of people and will cause disgruntlement or jealousy. In truth, public-sector salary data is public information in most jurisdictions, and a surprising number of public-sector employees can quote the salaries of their co-workers with some precision.

Project cost accounting requires making an investment in systems or processes. But, if sophisticated project cost accounting systems are not worth the investment, proxies for actual cost data can be used. Instead of charging actual salaries to projects, a single hourly rate or bands of rates can be used. The project cost data, obviously, will not be completely accurate, but it will be better than not having any cost data.

PROJECT SELECTION AND PRIORITIZATION

Before initiating a project, the organization should identify which projects it wants to invest in and which ones it does not want to invest in. Too often, however, senior public-sector managers want to have things both ways. They want the ability to constantly add new projects while not compromising performance on the existing projects. They want hard-pressed staff to do both this and that, and they do not often engage in resolution of the tough questions of project prioritization. Unfortunately, when everything is a high priority, nothing is a priority and nothing gets done. As long as we live in a world of constrained resources, priority is a concept that can only be applied relatively, in that the priority of a project can only be defined in relation to other projects.

The same problem exists with project selection. Although some public-sector organizations use excellent project selection methods, many do not, and projects simply get added to the project pile without any consideration of the relative merit of those projects. To be sure, some public-sector agencies are buffeted by must-do projects that are embedded in legislation, rule, or the mandates of higher-level agencies and jurisdictions. For example, federal Medicaid managers may require that states change their systems by a certain date. In some cases, public-sector projects are created by emergencies, including natural disasters and security threats.

Nonetheless, project priorities have to be set, and the projects that support the public interest best have to be selected over those with less public impact. In order to make informed decisions, public-sector managers have to deploy tools for project selection and prioritization. In an ideal world, projects could be selected on a purely quantitative, financial basis. Those methods can be applied in the private sector if costs and benefits of projects can be identified. Costs are fairly easy to estimate. Most project costs are knowable up front, and with careful estimating, reasonable cost estimates can be developed and used to calculate the net benefit of projects.

Even in the private sector, however, estimating project benefits can be more of an art than a science. When a project's benefits include efficiency gains, how are those gains to be measured? We can estimate the amount of time that employees will save because of a new process. But will those employees make good use of that saved time or remain underemployed? Will we be able to reduce the number of workers engaged in the process? If so, how sure can we be that those workers or their managers will not just find other work to fill their time? If they do, can we still attribute cost savings to the project?

With diligence and solid rules and guidance, most private-sector organizations can reasonably quantify the costs and benefits of their projects, and, by applying solid mathematical analyses, they can compare dissimilar projects against one another.[1] Some public-sector projects lend themselves well to mathematical justification. For example, if a development project were intended to attract a specific business to a community, we could identify the benefits of that project, which might include:

- The taxes paid on income directly related to that business
- Indirect benefits, including the taxes to be paid by those who hold jobs indirectly related to the business, like restaurants and grocery stores that would benefit from new economic power of the employees of the business
- Reduced social-welfare costs related to the creation of employment

Costs of the project would include:

- Foregone taxes related to incentive packages
- The costs of grants or other incentives
- The investment of the time of the development officers

In theory, we could, therefore, calculate the net present value of that project and decide whether or not we should undertake it.

In practice, however, public-sector projects—even those like the one above that allow some cost and benefit estimation—are a challenge. Although costs can be estimated, the benefits of public-sector projects often extend to the public at large or key stakeholders and defy quantification. As was indicated earlier, stakeholders for public-sector projects often include future generations, and the impact on those future generations is nearly impossible to determine.

For example, should the project to attract business to a political jurisdiction described earlier be afforded a higher priority if the

---

[1] The most theoretically sound method of comparing the costs and the benefit of projects is net present value (NPV). That method discounts future flows of costs and benefits, using a discount factor that incorporates the risk factors related to future funds flows, to a common point in time (i.e., the present), and calculates the difference between benefits and costs. A close companion, though harder to calculate than NPV, is internal rate of return, which establishes a discount rate for projects that becomes a threshold for project selection. There are many solid sources of instruction in the application of both techniques, and they will not be discussed further here.

business is located in the district of a member of the legislature of the same party as the administration who is vulnerable to challengers? Should it be afforded a higher priority if it is in the home district of the Speaker of the House? What if a major newspaper has highlighted the issue and is urging that the government do something to capture that business and those jobs? Although we might argue that these factors should not influence the prioritization of public-sector projects, anyone who has been involved in public-sector projects can attest to the fact that factors like these do influence project priorities. One could also argue that, because identifying the public interest is an imperfect process, these types of factors that influence public-sector decisions are appropriate and useful indicators of need.

Rather than completely abandoning methods that apply quantitative analysis to project prioritization and project selection, some public-sector organizations have successfully adopted mixed methods of project evaluation. These mixed methods use financial analysis as *one* criterion among many that can be used to evaluate a project. For example, a public-sector organization could construct an evaluation tool like the one illustrated in Table 7.1.

In this mixed model, there is an attempt to apply solid evaluation and use return on investment (ROI) as one of the criteria. It must also apply criteria that are realistic to the circumstances of the agency. Agencies developing such a method for the prioritization and selection of projects can use the criteria that are most relevant to them and weight those criteria in a way that recognizes the relative importance of those variables. For best results, independent persons should score each project, or a consensus score should be assigned to each criteria.

REQUIRED FUNCTIONS FOR MANAGING PUBLIC-SECTOR PROJECT COSTS

Managing public-sector project costs requires the performance of three critical functions. They are:

1. Estimating the costs of the project
2. Acquiring the financial resources for the project
3. Managing project costs and reporting on expenditures

Each of these functions will be considered in turn.

Table 7.1 Project Selection and Prioritization Matrix[2]

Criterion	Weighting	Score
Return on investment (internal rate of return)[a]	20–25% — 25 points 15–20% — 20 points 10–15% — 15 points 5–10% — 10 points 0–5% — 5 points less than 0% — −10 points	
Impact of the project on citizens	High — 30 points Medium — 20 points Low — 10 points	
Impact of the project on the strategic goals of the enterprise	High — 20 points Medium — 10 points Low — 0 points	
Probability of project success, which incorporates all project risk factors	High — 20 points Medium — 10 points Low — 0 points	
Statutorily mandated	Yes — 50 points No — 0 points	
Degree of political interest in the project (numbers of political persons with interest in the project and position of persons interested in the project)	High — 20 points Medium — 15 points Low — 0 points	
Impact on agency/program funding (e.g., will project failure limit access to federal funs	High — 30 points Medium — 20 points Low — 10 points	
Potential of project failure or nonselection to increase legal liabilities (e.g., ADA noncompliance)	High — 30 points Medium — 20 points Low — 10 points	
Other factors	30–0 points	

[a]The internal rate of return (IRR) on an investment or project is the discount rate at which the NPV on the project is zero. A higher IRR is, obviously, preferable to a lower one, and projects can be evaluated based on a hurdle or threshold rate.

[2] There are a variety of methods, including the use of criteria matrixes and value weighting, that can be used to score the relative importance to the organization of each criterion.

Estimating the Costs of the Project

Some public-sector projects are undertaken irrespective of the costs of the project. When the public safety is at stake, projects are undertaken with the intention of identifying costs later. For example, when a nation is attacked, no one attempts to determine if national defense would be cost effective. Most projects, however, are undertaken with enough forethought to allow for the estimation of the costs of the project in advance. Project cost estimating identifies the total cost of the project, and cost estimates are based on what is known about the project at the time of the estimate. Obviously, that estimate will change and become more accurate as the project team employs progressive elaboration of the project and learns more about it.

At the beginning of the project, all that can be accomplished is a rough-order-of-magnitude estimate. That estimate is used to provide a very general idea of the project costs for the purposes of selection, prioritization, and resource assignment. Rough-order-of-magnitude estimates are very inaccurate. For example, the Big Dig project, which was undertaken to sink Boston's central highway artery leading through the city below street level, was originally estimated to cost $6 billion. Total project costs were $14.6 billion, making it the most expensive public-works project in U.S. history.

The most common tool for initial, rough-order-of-magnitude estimating is analogous estimating, which simply requires that we find another project similar to ours that was undertaken in the past by a sister agency and modify that cost estimate by any known factors. Another agency might, for example, have conducted a project five years ago to deploy a new accounting system. We can take the cost of that project and apply inflation factors to it to determine the budget for the deployment of a new system for our agency. Obviously, this analogous estimate overlooks several factors, including changes in technology, the ability of staff, changes in hardware needs, and others.

As we get further along in the progressive elaboration process, we can apply parametric estimates, which require that we learn enough about the project to apply a mathematical formula to project estimating. For example, we might be able to identify the number of lines of code that will be necessary for a public-sector accounting application. If we can find a standard for the number of lines of code that can be written in a day, we can calculate the length of the project and, by extension, the cost.

Ultimately, once we have identified the activities necessary for the project, we can do bottom-up estimating. Bottom-up estimating usually applies expert judgment to each activity to estimate its cost. Bottom-up estimating is regarded as the most accurate estimating technique, although a bias applied to each activity can multiply across the activities to create a wildly inaccurate estimate of total costs. An even better strategy would be to compare the results of bottom-up estimating to the results of analogous or parametric estimating to see if the bottom-up estimate is within the realm of reasonableness. Note that these are the same estimating techniques that were used in estimating the length of the project.

In public-sector projects, we are often handicapped by the establishment of a fixed budget early in the project. For example, legislative appropriations may detail project budgets and dates by which programs must be established. The result is that project scope is the only leg of the triple-constraint model that we can adjust when problems occur. In the end, project cost estimates can be used to evaluate competing projects, manage project performance, and acquire the resources necessary for it.

If we want to apply earned-value management, which will be described later in this chapter, we must assign project costs to individual activities. Assigning costs to activities is a process that has already been completed if we performed bottom-up estimating. In that case, we need to sum those costs upward to create the total project budget.

If we used a top-down estimating method, we now need to allocate those costs to individual activities. Several methods can be used to allocate costs to activities. The most common method would be to allocate costs on the basis of the work units required for each activity. For example, if the total project required 100,000 hours of work and a total budget of $3 million, we could allocate X dollars of cost for every hour that an activity were estimated to take.

Cost identification and allocation systems are required for some public-sector programs, particularly those that are co-funded by other levels of government. For example, U.S. universities are required by federal programs to identify the full costs of operations, including indirect costs, and create a mechanism to allocate those costs to research projects that receive federal funds. The amount billed to the federal government becomes, therefore, a combination of direct costs and an indirect cost allocation percentage that is approved by the government.

Acquiring the Financial Resources for the Project

Acquiring resources for public-sector projects requires a combination of the ability to create a solid business case for the project and the ability to engage in formal or informal organizational politics. In some public-sector projects, the project budget has to be integrated into the agency's budget request. That will require that the project team engage in a variety of activities that include:

- Discussing budget processes with agency fiscal staff
- Completing budget request forms
- Defending the budget proposal to those who must provide approval
- Reporting on project costs
- Working with budget officials to identify and secure ongoing funding needs, including requesting additional funds, participating in future budget processes, and requesting changes in budget authorizations

In order to compete with other projects for scarce funds and get budget approval for projects at the agency level or higher, the project team will need to create a good business case for it. That business case combines many of the elements contained within the project charter, coupled with cost-benefit analysis that justifies the project and identifies benefits. The format for a business case to be used in the budget process will likely be prescribed by the agency.

Public-sector projects are rarely concerned with cash flows. As long as budget authority exists for the project, it is usually assumed that someone else will take care of meeting cash needs. That budget authority, however, does create problems in public-sector projects. Public budgets are created on strict and limited time cycles, and projects cannot assume that a budget started in one budget period will extend into another. Most governments operate on the principle that no budget bill can obligate a future legislature, which limits the length of budget authorizations for projects. A long-term project, such as a major construction project or information systems installation, for example, will extend across several budget cycles and must be appropriated in pieces.

Budget processes may, fortunately, contain mechanisms for mid-budget funding shifts employing a subset of the legislative body. For example, agencies may be allowed to request the authority to move funds from one line item to another. However, if a project is explicitly enabled by a budget statute and the need for additional funds is identified, the source of the additional funds will also need to be identified and those mechanisms engaged. Projects that are not explicitly budgeted by statute provide the agency with more flexibility, and the agency may have the ability to shift funds among projects without legislative approval. In any case, if more funds for projects are to be acquired, those funds need to be justified in a dialogue with senior managers who can budget for those funds.

Although some mechanisms exist for managing project budgets in the transition from one budget to another, projects are often broken down into time periods that are arbitrary from the perspective of the project. Public-sector project managers have, therefore, the additional responsibility of arguing for extension of projects into the next budget period and estimating the amount of funds that will be required in that upcoming period. This is particularly complicated with projects that have begun in one budget period but may not be completed for several periods. In that case, a budget estimate is required, not for enough funds to complete the project, but merely for enough funds to fuel one year's worth of performance. Some jurisdictions employ longer budget cycles, such as biennial budgets, which smooth the problems of project budgeting. Those same budget limitations may constrain the ability of the project manager to engage in long-term contracts with vendors.

In some public-sector organizations, a budgeting process will be undertaken inside the agency. In others, informal decisions on project selection are made at the agency level, and the formal budget process works at an aggregate level. If the agency-level decisions are informal, project sponsors and project managers must become adept at understanding those decision processes and the stakeholders who influence them.

The desired results of the process involved in acquiring the financial resources for the project are adequate funding for the project and support for it. The project manager will need to apply sophisticated organizational and communications skills in order to secure that funding commitment. In order to protect that funding and integrate the project's funding with budget cycles, the project manager must also be diligent in monitoring the funding and budget environment to identify threats to project funds. As budgets are cut—an increasingly frequent occurrence in the public sector these days—project managers must remain willing and ready to redefend their projects and compete with other compelling uses of the funds.

Managing Project Costs and Reporting on Expenditures

Like the other project management functions described thus far, cost control in public-sector projects can be more complicated than in the private sector. At least four factors complicate public-sector project cost management:

1. Project managers in the public sector may have less control than private-sector project managers over the composition of their project teams and over the amount of compensation the team members receive. Rates of pay in the public sector are often driven by years-of-service or time-in-grade criteria rather than project contribution. If the project manager is assigned a team of senior employees, project costs can escalate without a corresponding increase in performance.
2. Public-sector benefit systems can cause large and lump-sum impacts on project budgets. In some cases, an employee who becomes ill can charge large amounts of pre-earned sick leave to the project budget. Senior public-sector employees also have accrued large amounts of vacation time, which in some cases will be charged against the project when that vacation is taken, which has both a cost and performance impact on the project.
3. Some public-sector projects are charged overhead rates that charge the costs of general government to projects and chargeable activities. In publicly supported universities, overhead rates can exceed 50 percent, which causes $1 of direct project cost to incur a total of $1.50 in project charges.
4. Public-sector purchasing systems can limit the ability of the project manager to control costs. If items are purchased for the project, it is likely that the agency's purchasing unit will be assigned to bid and purchase the items. Those processes are often slow and may not allow for the most cost-effective purchases to be made. If vendors are employed in projects, the same purchasing systems can slow the contracting process and add unnecessary costs to the project. For example, set-asides for minority vendors may require that a minority vendor with higher costs be selected over a non-minority vendor with lower costs. Although those set-aside programs accomplish laudable public goals, they have a cost impact on projects.

Those factors do not imply that public-sector project management should not be held accountable for cost management. Instead, a

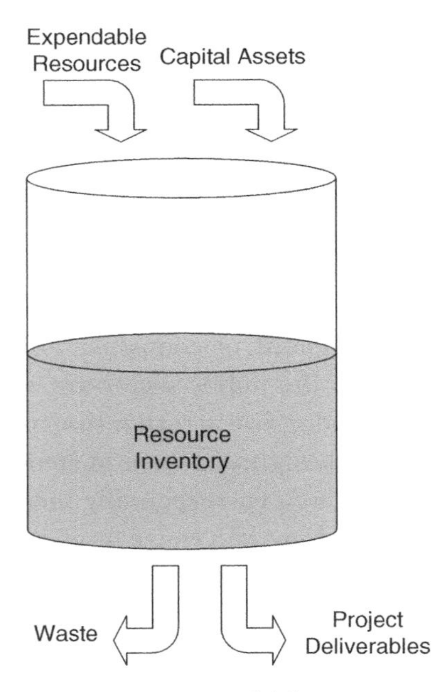

Figure 7.1 Input-Output Model for Project Resources

public-sector project manager has to recognize the limits on his or her ability to manage costs, take those into account, and manage the project's costs to the best of his or her ability. In addition to ensuring that project costs do not exceed budgeted levels, project cost management also includes the activities necessary to optimize the use of project resources. Figure 7.1 illustrates a simple input-output model for project resources.

First, at the top, this model identifies the flow of resources into the project that allows it to do its work. These assets are the inputs to work processes. There are two major types of assets/inputs available to the project. They are expendable, assigned assets and facilities and capital assets, as detailed in the following sections.

Expendable, Assigned Assets

Expendable assets assigned to the project create direct costs that can be deployed by the project to accomplish its work. Those assets include staff and supplies, and their assignment to work activities creates direct costs. The costs of these resources are the most controllable type of costs, and most direct costs represent the costs of resources assigned to the project.

Direct costs can also be variable costs, in that they can be expended based on the level of output.

Assignment or expenditure of some assets kicks in cost multipliers. Labor costs cause the most cost multipliers. An hour of direct labor has a direct cost based on the hourly rate or salary of the individual. Added to that, however, are the costs of health-care benefits, the costs of vacation and sick leave, and the costs of any other benefits. Because many indirect costs are assigned based on labor charges, the deployment of labor can generate higher indirect costs for the project as well. The project can also argue for the assignment of more of these assets if resources are inadequate.

Facilities and Capital Assets

Facilities and capital assets are also assigned to the project to allow it to accomplish its goals. They include equipment and other physical assets, including the space assigned to the project. These assets do not represent variable costs, in that their use cannot be varied dependent on the level of services. For example, a piece of equipment has a fixed price irrespective of the number of users of the system.

In order to make the best use of facilities and capital assets the project needs to optimize their use. As capital assets are used, their costs are capitalized and assigned to the project over time through depreciation or amortization. The combination of these resource in-flows creates an inventory level of unit capability. Inventory has a holding cost, because inventory of capability held by one project cannot be used by another. One special type of waste is excess unit capacity. Excess capacity builds the inventory level of resources without an accompanying increase in goal attainment.

The utilization of these assets is also represented in Figure 7.1. As the project performs its duties, those assets are used, and the costs related to those assets are incurred. The goal of the project is to optimize the use of those resources. Resources can be consumed in two ways that are described as follows.

Creation of Project Deliverables

Some (we would hope, most) of those assets can be expended on creating the deliverables of the project. In order to assess the extent to which these assets have been optimally deployed, the project team has to clearly understand what its deliverables are and has to be able to measure the attainment of the deliverables.

Waste

Not all of the assets of the project can be utilized effectively. When they are not deployed and used optimally to create the project's deliverables, they become waste. Waste represents the difference between actual resource use and optimal use. Waste can include such items as unnecessary overtime, unused or poorly used supplies, downtime, overstaffing, and waiting time. The goal of project management should be to reduce waste. Note that as more assets are provided to the project, the opportunity for incurring waste also increases.

In order to report on project expenditures, public-sector project managers must first employ cost reporting systems that accumulate costs. As indicated earlier, in some public-sector project environments, cost reporting systems are lacking. If the project manager cannot convince the agency of the need for good cost accumulation and reporting, the project manager can consider creating his or her own cost reporting system. These systems can be informal and can use proxies for detailed cost items. For example, because time spent on projects is usually the greatest cost factor, project managers can create a standard hourly or daily rate for project staff and develop approximations of actual costs.

That cost data can be used to help the project manager measure project accomplishment and to hold staff accountable for outcomes within agreed-to amounts. That is, staff, when assigned project activities to be accomplished, can also be assigned a number of hours in which to complete that assignment. This data can help the project manager determine if activities have not been achieved because of other demands on the team member that limited his or her ability to work on the project or because of the team member's inefficiency. The data can also help the project manager determine which team members deliver the best value to the project and provide an estimate that the functional manager can use to determine how much of his or her resources will be given up to the project.

If project cost reporting to senior managers is required, the project manager should keep in mind that manager's understanding of the project and his or her understanding of cost factors. Summary data and data interpretations may be required.

EARNED-VALUE MANAGEMENT OF PUBLIC-SECTOR PROJECTS

Even if we construct a project schedule and a project budget, project managers are faced with a challenge when they assemble standard project performance information. Typically, they can only tell how much they have spent to date. But what does that tell them? If they have not spent as much as they had expected, does that mean they are under budget and should throw a party? Or does it mean that the project just got a late start, in which case they are behind schedule?

Earned-value management presents an elegant mechanism for answering those questions and linking cost and schedule data with data on how much work has been accomplished. To understand the application of earned-value management to public-sector projects, a short overview of earned-value management may be necessary.

Earned-value management begins with the creation of three new terms:

- *Earned value* (EV) is the budgeted cost of the work that has been performed to date.
- *Planned value* (PV) is the budgeted cost of the work that has been planned to be performed to date.
- *Actual cost* (AC) is the actual cost of the work accomplished.

For example, assume that the project consists of 100 activities arrayed over a six-month project. At the end of the third month, we want to know how we are doing. If we add up the budgets assigned to all of the activities accomplished so far, we can determine the EV. If we add up all of the activities we had planned to accomplish by this point, we can calculate the PV. If we add up the costs of the activities we have actually performed so far, we can calculate the AC.

Four very simple formulas allow us to determine how we are doing. They are:

- $EV \div AC$, which calculates the *cost-performance index* (CPI)
- $EV - AC$, which calculates the *cost variance* (CV)
- $EV \div PV$, which calculates the *schedule-performance index* (SPI)
- $EV - PV$, which calculates the *schedule variance* (SV)

To identify how we are doing with our project, we actually need to apply only two of the formulas, because both cost terms tell us the same thing, and both schedule terms do the same.

For the index terms (CPI and SPI), a result of greater than 1.0 indicates good performance. A CPI of 1.08, for example, indicates that we are getting $1.08 of budget performance for every $1.00 we invest in the project. A result of less than one indicates poor performance. An index of exactly 1.0 indicates that the project is right on schedule or budget, depending on which index has been calculated.

For the variance terms (CV and SV), a positive number indicates good performance. For example, an SV of +$7600 indicates that the project is ahead of schedule. The reverse is true of a negative result, and a result of 0 indicates that the project is right on schedule or budget.

These calculations are simple and supported by the software. A minor complication occurs for activities that, at the point of the analysis, have been started but not completed. A variety of rules can be applied. The most conservative gives to earned-value (EV) "credit" for an activity that has been started but not completed. Another rule is the 50/50 rule, which provides for half of the activity's earned value when it starts and the remainder when it is finished. The 25/75 rule works the same but is a little more conservative.

We can also use the CPI to estimate the total cost of the project given the performance to date. That new estimate is called the *estimate at completion* (EAC). We can use three different assumptions to make those estimates:

- We can assume that the variances encountered to date will continue through the rest of the project by dividing the total original budget, which is referred to as the budget at completion (BAC), by the CPI (EAC = BAC ÷ CPI).
- We can assume that the variances found so far will not continue by taking the costs to date and adding the budgeted costs of the work that is left (EAC = AC + (BAC − EV).
- We can start over and create an entirely new cost estimate by adding the actual costs incurred so far to a new estimate of the costs required for completion (ETC) (EAC = AC + ETC).

Similar calculations do not work for schedule estimates.

BEST PRACTICES IN PUBLIC-SECTOR PROJECT COST MANAGEMENT

In order to establish budgets for public-sector projects and manage those costs, best practices include:

- Estimate project costs at the lowest level of project detail
- Keep in mind, to the extent possible, that project budgets will change as more detail is known
- Urge the assignment of resource costs to project activities through timekeeping systems, employing proxies or standard rates if actual costs are not available
- Use earned-value performance management to track project status
- Work with budget officers and staff to incorporate project needs into budget documents
- Remain aware of and factor overhead rates and multipliers into project costs
- Build a compelling business case for the project that can be used to secure resources and defend the project from later cuts
- Factor benefit payouts into project risks and work with human resource and accounting staff to shift the burden of benefit payouts away from the project to the extent that the benefits were not earned during project work
- Build cooperative relationships with purchasing agencies in order to win the right to participate in purchasing decisions to the extent allowed by rule
- Be prepared to argue for the funding of your project at its initiation and throughout as budget cuts are made
- Think about optimizing the application of resource inputs to create the most deliverables
- Reduce asset inventory levels

DISCUSSION QUESTIONS

1. What challenges have you encountered in identifying project costs or applying cost management in your public-sector projects?

(continued)

2. What challenges have you had in managing projects across budget cycles? What tools have you used to make those transitions across budgets?
3. What challenges have you experienced in estimating costs for your projects? What cost-estimating methods have you used? How accurate have your estimates been?
4. What cost management challenges have you faced in your projects? Do public-sector accounting and purchasing processes add costs to your projects? What other factors increase your project costs?
5. What other best practices can you imagine for managing project costs in public-sector projects?

EXERCISES

1. For a project, you have identified the following:
   - You have completed 32 tasks to date with a combined budgeted cost of $24,000.
   - You had planned by this point to have completed 34 activities with a combined budget of $26,000.
   - The actual cost of the activities you completed to date is $23,000.

   What conclusions can you make about the project?

   How would the analysis change if you had also begun another six activities by today, which had a combined budget of $4,000? Assume that you are using the 50/50 rule.
2. Assume that:
   - The original project budget was $40,000.
   - The amount spent to date is $10,000.
   - The CPI is 1.2.

   Given that the CPI is above 1.0, will the new estimate of the project's cost be higher or lower than the original $40,000? What is the EV in this case? Calculate the new estimate assuming that the variances will continue.

   Now assume that we have spent the $10,000 to date but still have activities with a combined budget of $32,000 left to perform. Calculate the EAC assuming that the variances to date will not continue. What is our EV in this case?

Turning on the Lights in the Country

In the early years of the 20th century, households and businesses in urban areas of the United States rapidly acquired electricity and enjoyed increasing standards of living. By the 1930s, 90 percent of urban families had electricity. Unfortunately, the same was not true of rural areas. Only 10 percent of rural families had electricity, and they paid rates that were commonly twice the urban rate. Electric utility companies argued that rural markets were not able to be served economically because of high network construction costs and little immediate return. More than 30 state initiatives were undertaken to bring electricity to rural consumers. With the advent of the Great Depression, most of those initiatives failed.

Some argued that the federal government had no right to get involved in the delivery of electricity to rural areas, because they would compete against private companies and because those initiatives would take the nation too close to socialism. President Franklin Roosevelt, however, believed that it was the government's duty to provide electricity to the people if private companies could not. As a result, he signed Executive Order 7037 in May 1935, creating the Rural Electrification Administration (REA), with the goal of extending the benefits of electrification to rural areas of the country. A year later, the Congress passed a bill authorizing $410 million for a 10-year program to electrify America's farms.

The REA adopted a cooperative model that had been successful in Pennsylvania. Those cooperatives were consumer-owned firms managed by a board of directors elected by the members of the cooperative. The REA, in turn, made loans to the cooperatives at low interest rates and with a generous 25-year repayment schedule. Provisions were built into the program to ensure that the cooperatives did not compete with private companies, and incentives were provided for connection to the existing electric network. Those loans to rural electric cooperatives became one of the largest investment projects of the New Deal. By 1939, the REA had helped establish 417 rural electric cooperatives, serving 288,000 households.

The REA is considered by some to be an example of government policy making at its best. By the mid-1950s, nearly all U.S. farms

(continued)

were electrified, and the default rate on the loans was less than 1 percent. In addition to simply providing electricity, the REA increased rural standards of living and created new consumers for U.S. manufacturers, as rural consumers began to buy washing machines and other appliances. It also linked rural areas to the rest of the country through media like telephones and radios.

Chapter 8

Managing Project Quality

THE BASICS OF PROJECT QUALITY MANAGEMENT

Project quality management focuses on both the quality of the project itself and the quality of the products produced by the project. The quality of the project can be measured by such indicators as its timeliness and its adherence to its budget. The quality of the product produced has to be measured in relation to the uses to which the product will be put. It does no good to deliver a project on time and on budget but fail to deliver products that are useful for the agency. Project quality management, which sometimes gets forgotten in the rush to meet project deadlines, is intended to increase the probability that the project will deliver useful results and reduce the risk of dissatisfied stakeholders.

Project quality management can benefit from what has been learned by those engaged in improving the quality of processes. Pioneers in that movement, including Philip Crosby, Joseph Juran, and W. Edwards Deming, have taught that:

- Quality management is an ongoing process.
- Prevention of mistakes and errors is preferable and less costly than inspection and detection of errors after they have been built into products.
- Quality is a management responsibility; workers will meet quality standards only when they have been provided work systems and incentives that allow them to meet those standards.
- Quality of the product is a matter of satisfying customers and meeting their standards.

The cost of quality is a related concept that defines the costs of quality as being prevention, appraisal (or inspection), and failure. The modern quality movement suggests that the total cost of quality can be reduced if the emphasis is shifted from failure and inspection to prevention. Because the quality of the product is related to the satisfaction of stakeholders, one definition of project quality, which relates primarily to the product of the project, suggests that project quality is defined by the ability of the project's deliverables to meet the criteria demanded by the project sponsor and users of the deliverables. As explained later, identifying stakeholder requirements is a challenge that many projects fail to meet.

THE CHALLENGES OF PUBLIC-SECTOR PROJECT QUALITY MANAGEMENT

Too often, the phrase, "good enough for government work," is used. That adage implies that public-sector operations and projects cannot expect to deliver high-quality outcomes and that public-sector stakeholders are reconciled to receiving mediocre products. In truth, the reverse is more accurate. Public-sector stakeholders, including the press and the public, are very demanding and have little tolerance for failure. Because government services are paid for with taxes, all stakeholders feel that they have the ability to place demands on public agencies and expect performance tailored to their needs.

The toughest quality challenge for public-sector project managers is meeting those stakeholder demands with the limited resources available to agencies. Although the goal of senior managers in public agencies is to ensure that those agencies are funded with the minimum acceptable level of resources (thereby optimizing agency efficiency), agency project managers are often expected to create first-class results.

Other impediments to the achievement of quality standards in public-sector projects are embedded in the rules and processes levied on project managers. Those impediments to meeting quality standards can include:

- Hiring and retention systems that reward longevity and do not require employees to be well-versed in the most current technologies and methods
- Purchasing systems that require the low bidder rather than the most competent bidder to be hired (Fortunately, some public-sector

purchasing systems allow for the selection of vendors based on both cost and quality.)
- The inability to provide incentive payments for exemplary work
- Purchasing systems that do not make allowances for hiring or contracting with high-cost experts
- Changing project priorities caused by electoral cycles
- Aged infrastructure that project deliverables might be required to interface with

The activities required to manage project quality standards and identify tools that might help public-sector project managers deal with these daunting impediments will be examined.

THE FUNCTIONS REQUIRED FOR PUBLIC-SECTOR PROJECT QUALITY MANAGEMENT

Two required functions were identified for managing public-sector project quality:

- Identify project quality standards
- Manage project quality

Each function is described in the following sections.

Identifying Project Quality Standards

Quality planning allows the project manager and project team to identify the quality standards that will apply to their projects and build a plan for achieving them. The principle outputs of this process are the quality management plan and quality metrics for the project. The quality management plan and its associated quality metrics become part of the overall project management plan.

The quality management plan can contain such elements as:

- The metrics themselves
- The checklists for managing project quality
- The processes to be employed for managing project quality
- Processes for stakeholder review and acceptance of the product of the project

- Responsible persons
- Process improvement plans

Quality metrics for a public-sector project should:

- Contain both project and product metrics
- Be measurable using data that can be derived in a cost-effective manner
- Not contain any elements that are not within the scope of the project
- Contain understandable and objective criteria
- Be reasonable, because the project team has committed to their delivery within the constraints of the project
- Be agreed to by stakeholders and project teams

Too often, either quality metrics are not identified for public-sector projects or they are poorly identified. Some examples of poor public-sector project metrics are:

- To make the system easier to use
- To meet the requirements of the statute
- To satisfy stakeholders
- To deliver a product by a preset date (without identifying clearly what the product should be)

Better metrics for public projects would be:

- The system shall allow users to access system menus with no more than a three-second wait time.
- The new administrative rule shall contain clauses that describe the responsibilities of each level of government.
- The program shall be used by 95 percent of county treasurers within 90 days of its deployment.
- The process shall reduce the processing time for licenses by 50 percent on average.

Each project would, obviously, require a complete set of metrics that measure the attainment of both project and product outcomes. Metrics are intended to impact performance, and the use of a single metric or a limited set of metrics can impact performance in unintended ways. For example, imagine that telecommunications regulators wished to induce telecommunications companies to adequately staff their customer call centers. A metric that they might consider would be that *a customer call must be picked up within 20 seconds 98 percent of the time.*

In general, the metric could be regarded as a good one. It is measurable and does not require 100 percent attainment, which would be unrealistic. But, assuming that were the only metric applied, imagine what could happen if a call were placed on hold for 22 seconds and another call were received. Which call would the company have been provided an incentive to pick up first? The person on hold has already failed the metric; the new caller has not. As a result, the rational response with regard to the metric for the phone company would be to leave the first caller on hold and answer the second call. This does not presume that the companies would handle calls in this manner, but, clearly, the metric has provided them an incentive to do so. A simple fix would be to also require the company to answer calls in the order in which they were received.

Note also the use of the word "shall" in the metrics. "Shall" ensures that the metric is necessary. It is preferred over words like "should" or "can."

Managing Project Quality

Managing project quality requires the project team to undertake activities such as monitoring quality activities, improving processes for quality management, and performing quality audits. Because public-sector projects are constrained by rules and required processes, another set of activities related to quality management for those projects includes those related to conformance with rules and standards. That can include analysis of whether the project has complied with those rules and processes. In some cases, those types of reviews are undertaken by oversight agencies and may take the form of compliance audits. Because sanctions can be applied to noncompliance and because the public-sector environment is often confrontational, those compliance audits can be a source of stress and can be difficult to keep in perspective. It is important to remember that those rules and processes have been created with good intentions in mind. The tension between those responsible for enforcing conformance with the rules and those responsible for creating project results is natural and not personal. The goal should be to create a productive balance between the goals of the project and the legitimate needs of those responsible for enforcing rules.

Quality management also includes the monitoring and evaluation of specific project outputs to determine whether they have met quality requirements. It also includes the elimination of the causes of unsatisfactory results. Quality management is different from scope verification, a process examined earlier. Scope verification is that process that allows stakeholders to determine whether project results are *acceptable* and whether they will sign off on the receipt of those deliverables. Quality control measures whether project results are *correct* from the perspective of quality metrics. Correctness has a higher threshold than acceptance, and stakeholders may accept some results that are not completely correct.

A principal tool of quality management is the comparison of the outputs of the project with the planned outputs. It may require inspection of outputs and corrective action if those outputs are not up to quality standards. Quality control also requires an attempt to find out why the desired results were not obtained and correct any problems discovered.

Most public-sector projects create unique deliverables, and few are amenable to the kinds of quality control tools, like Total Quality Management and Six Sigma, that are used to improve processes. Nonetheless, public-sector projects should identify quality metrics and evaluate projects and their outcomes to determine whether those metrics have been met with as much rigor as is cost effective.

Two tools for helping the public-sector project team improve the quality of their projects and the processes that those projects interface with are described as follows:

- Lean government
- Project requirements elicitation and management

Each will be discussed in turn.

LEAN GOVERNMENT AS A TOOL FOR QUALITY IMPROVEMENT

Many public-sector agencies are well-versed in the application of methods for the improvement of processes. Total Quality Management (TQM), Continuous Quality Improvement (CQI), and Six Sigma have been deployed by many agencies and made a part of the organizational culture.

Most recently, public-sector agencies have applied the concepts of Lean government to their processes. The Lean methods or principles, which have been used very effectively in the private sector to reduce inefficiencies in manufacturing and transaction processing, focus on the identification and elimination of non-value-added activities (defined as waste) involved in delivering products or services to customers. Some Lean advocates argue that only 5 percent of process activities actually add value for the customer, making 95 percent of activities either non-value-adding or waste. The Lean terminology initially derived from descriptions of the process-improvement and waste-reduction efforts employed by Toyota.

Waste in processes also correlates to the waste identified in the chapter on project cost management when the flow of resources and outputs. The principal types of waste identified in Lean methods are as follows. Their impacts on public-sector projects are also listed:

- Defects (incorrect deliverables)
- Overproduction (gold plating or creation of extra deliverables)
- Conveyance (shuffling deliverables between different parts of the organization)
- Waiting (the time spent by workers waiting for resources or inputs from other project team units or members)
- Inventory (deliverables that have not been transferred to users or excess capacity held by the project team)
- Motion (the movement of workers or supplies)
- Overprocessing (which can be created by providing services above those required or using overqualified resources to perform tasks)

The principal tools employed in the Lean government methods are value-stream mapping (VSM) and Lean events, sometimes called Kaizen events. VSM produces a set of visual depictions of the processes engaged in providing a service or product of value to the customer. By making the value stream visible, it identifies waste, delays, and the inefficiencies of organizational design. Kaizen events are intense workshops engaging an array of stakeholders in an effort to map and improve processes.

Advocates of Lean methods argue that they can:

- Eliminate backlogs, interruptions, and waste
- Cut costs
- Reduce lead times substantially
- Decrease process complexity

- Improve the quality of applications and the consistency of reviews and inspections
- Allocate staff to mission-critical work
- Improve staff morale and process transparency
- Allow effective process redesign

Six Sigma is a more complex, although related, process improvement tool. Lean methods can create similar results without the substantial investment of time and resources required by Six Sigma. Lean methods differ from TQM in that Lean methods not only address quality but also address cost and service delivery.

Lean methods are not about reducing the role of government, relaxing regulation, or privatizing services. They are a means of helping to move process-based organizations toward more flexible and goal-oriented ways of operation that provide a better fit with the environment in uncertain and demanding times. Although Lean methods were designed for and applied initially to manufacturing, they have evolved and been deployed in administrative processes and, more recently, to public-sector applications. The results cited by agencies that have applied Lean government strategies are dramatic. Some of those results were cited earlier in the discussion of projects designed to improve government processes.

MANAGING PROJECT REQUIREMENTS

The challenges of managing *project* quality for public-sector projects have been detailed as we described the management of project scope, time, and cost for public-sector projects, and will be further explored as we describe the other knowledge areas. Although those challenges are formidable, the challenges of managing the quality of the *products* of public-sector projects can be even tougher. And although we might have created a set of general quality metrics for the product of the project, we still need to create more detail about the specific requirements of the deliverables of the project. For example, although we might have identified the general requirements of a public-sector accounting system, we need to identify the specific requirements of the individual modules of the system, the user interfaces, and the system operations.

Managing requirements is always a challenge, no matter what the project environment, but identifying and managing the quality of the products of public-sector projects can also be compromised by:

- Legislative mandates that have not been clearly articulated or that did not anticipate a host of implementation challenges
- Political stakeholders who may be deliberately vague about their goals in order to escape criticism by some stakeholders
- The sheer number and variety of stakeholders for public-sector projects
- The requirements of public-sector processes, which may put reducing costs ahead of project and product quality
- The inability to secure capable resources on salaries paid to public-sector employees
- The challenges of motivating, rewarding, and penalizing public-sector employees, which can have a direct impact on the quality of deliverables

As a result, public-sector project managers may feel as if they are operating “in the dark” with regard to the specific needs of their stakeholders. In the information vacuum, they need to avoid the temptation to guess at those specific requirements.

In recent years, an increasing focus of project quality management has been on the identification of clear project requirements and the management of those requirements. Requirements define the attributes of the product of the project. This recent focus on project requirements was also driven by acknowledgment that most projects that fail do so because they fail to identify user requirements and deliver products that meet user needs. Good project management naturally embraces several tools for identifying and managing requirements. They include:

- A management philosophy of open management and constant dialogue with stakeholders
- Early identification in the project charter of the business need for the project
- Driving the project with solid scope definition
- Progressive elaboration and change management designed to accommodate changes in the project
- Scope verification
- Quality control
- Development of a communications plan that analyzes stakeholder needs and identifies ways to communicate with them frequently and effectively

Nonetheless, some project managers and business analysts felt the need to develop stronger methods and processes for identifying and managing project requirements. They formed the International Institute for Business Analysis (IIBA®) to serve as the professional association of business analysts, which they define as being the person responsible for project requirements.

The IIBA® also developed a standard for identifying and managing project requirements. They called that standard, which is the first standard created for the purpose, the *Business Analysis Body of Knowledge® (BABOK®)*. The *BABOK®* takes the important step of identifying a set of processes for requirements management called the requirements cycle. That cycle consists of:[1]

- Requirements planning
- Enterprise analysis
- Requirements elicitation
- Requirements documentation and analysis
- Requirements communication
- Solution assessment and validation

That requirements cycle fits within the project plan in that project requirements can be defined either as part of the project management plan or as a project-based deliverable of the project. And although the project manager is responsible for the overall project, including requirements, often a person is assigned responsibility for the identification and management of requirements. That person is sometimes referred to as the business analyst for the project.

This requirements process implies that three essential documents need to be produced if requirements are to be well-managed and well-understood. They are:

1. *The requirements management plan*: which describes how requirements will be elicited, documented, validated, and communicated
2. *The requirements list*: which is a list of requirements listing attributes like the number, priority, and source of the requirement

[1] International Institute of Business Analysis, *Business Analysis Body of Knowledge® (BABOK®)*, Version 1.6. That requirements cycle has been replaced in Version 2.0 of the *BABOK®*.

3. *The requirements package*: which contains the requirements list, diagrams, and other methods of analyzing those requirements

Two elements of the requirements cycle are especially critical to public-sector projects. The first is requirements elicitation. In early versions of the *BABOK®*, the term requirements "gathering" was used. Requirements gathering, however, does not do justice to the level of rigor required for creating a list of requirements. In practice, the process of identifying the needs of users is a time-consuming and challenging process.

Users of the eventual product of the project often do not want to take the time to share their needs. Sometimes they do not have a clear idea of what those needs are. And often, their requirements will change during the course of the project. In order to adequately elicit the requirements that will be demanded by users of the product of the project, the *BABOK®* identifies several requirements elicitation methods, which include methods such as:[2]

- Interviews
- Brainstorming
- Surveys
- Focus groups
- Prototyping
- Job shadowing
- Document analysis

The project manager and business analyst need to consider which of these methods of elicitation are most appropriate for their project. Project managers and business analysts are also aware that, in some cases, project stakeholders are invited to participate and help in the identification of requirements in order to let them feel a sense of participation in the project. That is an important function that does not, necessarily, contribute to effective and efficient requirements elicitation.

The second process in the requirements cycle that is useful to review is requirements documentation and analysis. Documentation and analysis operates on the premise that a requirements list usually does not

[2] Ibid. p. 155.

provide enough guidance for those who will develop the product of the project. They need more detail, including diagrams and charts to help them visualize the solution that is required by the customers of the project.

For analyzing and documenting project requirements, the *BABOK*® has identified a variety of methods. They include such methods as:[3]

- The analysis of business rules
- Process flow analysis, including
  - Activity diagrams
  - Work-flow diagrams
  - Data-flow diagrams
  - Event mapping
  - State diagrams
- Use cases and use case diagrams

These methods are not appropriate to every project. The goal of the application of these methods is simply to make certain that the requirements are complete and understandable, both to the customers who need to understand the scope of the project and to those who are responsible for building the product of the project.

Without good project requirements definition and management, the project can easily evolve into a series of conflicts with stakeholders and scope changes that can inundate the project and turn it into a set of ad hoc reactions to stakeholder needs. In addition, good project change management cannot occur unless all of the stakeholders of the project have a very clear, unambiguous definition of what the project is and what it is not.

Creating good requirements requires a dialogue with users, something that takes time and energy and may force project managers outside of their comfort zone. Although project teams may not want to engage in that dialogue, the days of satisfying users of the products of the project with an attitude of "if we build it, they will love it" is over. Users and stakeholders today demand and deserve a seat at the table as we define the products we intend to create for them.

---

[3] Ibid. vii–viii.

BEST PRACTICES IN THE MANAGEMENT OF QUALITY IN PUBLIC-SECTOR PROJECTS

Best practices in the management of quality in public-sector project include the following:

- Engage in a dialogue with project stakeholders about what it costs to achieve quality goals, and ensure that they have realistic expectations with regard to what can be achieved with the resources committed to the project
- Identify the quality impact of hiring and purchasing processes, and work with process owners to reduce negative impacts
- Identify the criteria that will determine project success, including project quality and the quality of outcomes
- Create metrics that meet the criteria identified earlier
- Include conformance to legal requirements and those requirements created by rules and processes as part of the quality metrics for the project
- Consider the use of Lean methods for improving processes
- Employ a standard requirements cycle designed for the needs of your organization
- Use "shall" statements to define requirements
- Ensure that project stakeholders understand the requirements elicitation and management processes and their roles in them
- Verify that requirements are complete and understandable by using the techniques for requirements analysis and documentation
- Deploy a requirements management process that requires frequent interaction and dialogue with stakeholders

DISCUSSION QUESTIONS

1. What challenges have you encountered in achieving public-sector project quality objectives?
2. How do you identify project quality goals? What goals have you identified? Who is involved in these decisions?
3. What Lean methods have you employed? What has been your experience?

(continued)

4. Which of these methods have you employed for requirements elicitation for your projects? How well have they worked?
5. How do you document and communicate project requirements? How do you ensure agreement on requirements?
6. What other best practices can you imagine for managing project quality in public-sector organizations?

EXERCISES

1. Design a set of quality metrics for a project. Make sure that the metrics meet the criteria listed in this chapter and address both product and project metrics.
2. For a project that you are familiar with, identify a set of requirements. Use "shall" to define them, and make certain that they meet the criteria for good requirements. How do these requirements differ from the answers to the preceding question?
3. Use one of the methods described to analyze and document the requirements you identified in the last exercise.

The Allied D-Day Invasion of June 1944

Managing public-sector projects can be difficult enough, but imagine managing a project involving nearly 200,000 team members, 5,000 ships, planes, and other equipment *and* keeping the project a secret.

That was the case with the Allied invasion of France in World War II. Although it was known that an invasion of France was being planned, it was critical that the enemy not know the time or the place of the landing. General Dwight Eisenhower, who could be regarded as the project manager, began planning for the invasion years before it occurred, and deliverables included assembly of the troops and logistics, a campaign to deceive the enemy, negotiations with the other nations involved, reporting to civilian leaders, and, ultimately, the implementation of the invasion and identification of lessons learned, which were facilitated by detailed after-action reports filed by the military units involved. The campaign to deceive the enemy included the creation of a fictional Army Group based across the English Channel from the supposed invasion site that included inflatable equipment and plywood and canvas installations. Those attempts at deception succeeded beyond expectations.

From the start, things began to go wrong and project changes were required. The original landing date was scheduled for June 5, 1944, but inclement weather did not allow it to take place. On June 6th, the weather was still poor, but General Eisenhower had to weigh the risk of landing in poor weather against the risk that the invasion force would be discovered. He made the decision to launch the invasion that day. He made a visit to say good-bye to the members of the airborne forces who were to land behind enemy lines and who were expected to incur up to 80 percent casualty rates.

On the ground, things also did not go according to plan, and countless improvisations were made at all levels of the chain of command. Gliders missed their drop points, and airborne troops were scattered. High tides on Utah Beach swept the landing craft away from their assigned landings. (Ironically, those high tides swept them away from the point of highest resistance.) The poor weather also aided the invasion force, as the enemy leaders felt that the weather was too bad for an invasion. Although chaos was the order of the day early on D-Day, order gradually was re-created as plans were revised and units took corrective actions. By the close of D-Day, the invasion forces had not achieved their full objectives, but they had established a hold on France. In that first day, more than 100,000 soldiers and sailors had landed in France by sea or air.

The D-Day invasion, although it failed in some respects to meet its immediate goals, was a success due to detailed planning, clear communications, effective change management, the support of civilian leadership, and, most importantly, the courage, devotion to the mission, willingness to endure hardships, and the personal sacrifices of the units and individuals involved.

Chapter 9

Managing Project Human Resources

Human resources represent the largest single cost item for most public-sector projects and some of the largest challenges. Managing those resources is about maximizing the contribution of agency and vendor staff assigned to the project. For some project managers, particularly those who are uncomfortable with conflict or dealing with people, human resource management is a special challenge.

THE CHALLENGES OF HUMAN RESOURCE MANAGEMENT IN PUBLIC-SECTOR PROJECTS

Human resource management in public-sector projects may differ more from private-sector projects than other knowledge areas. Public-sector project managers face additional challenges related to the management of their project teams that include:

- The inability to clearly link performance and rewards
- The fact that public-sector compensation systems are biased toward longevity, which means that the most productive and useful project team members may not be the most highly compensated
- The inability of the project manager to select project team members based on expertise (most public-sector teams are composed of existing employees in the unit)
- A culture of work performance that does not reward risk taking or the contribution of above-and-beyond performance at project crunch times

- Compensation systems that do not allow the retention of the highest performers with the most critical skills
- A workplace that, in some cases, is not familiar with results-oriented project management
- Political interference in the management of projects and the challenges of dealing with political appointees

In addition, most public-sector employees are not defined as "at will" employees, who serve at the pleasure of their employer. Typically, only those public-sector employees regarded as political appointees serve "at the pleasure." When employees are not "at will," they must be subject to progressive discipline in order to remove them, and that progressive discipline process must be conducted with the intention of improving the employee's performance. The progressive discipline process often proceeds from verbal warnings and counseling, through written warnings, to a series of increasingly long suspensions, and finally to termination. The result is that managing public-sector employees can take more time in the public sector than the private sector and usually involves interactions with human resource administrators and can involve unions and attorneys. These processes, though seemingly burdensome, are also intended to ensure that public-sector agencies do not become overly politicized.

This is not to imply that most public-sector employees are not hard working and conscientious. There are many factors that motivate public-sector employees, including the desire to be professional and advance their careers, their allegiance to their agencies and smaller units, and their concern for the public interest. The factors listed previously do, however, indicate that hiring, retaining, and motivating project teams in the public sector comes with special challenges.

THE REQUIRED FUNCTIONS FOR PUBLIC-SECTOR HUMAN RESOURCE PROJECT MANAGEMENT

We identified several required functions for managing human resources in public-sector project management. They were:

- Creating a plan for optimizing the use of human resources
- Motivating and managing the project team
- Resolving project conflict

These functions will be examined in turn. Management theories that apply to project management as well as the role of leadership in the public sector and its relationship to project management will also be discussed.

Creating a Plan for Optimizing the Use of Human Resources

The major cost item for public-sector projects is usually the cost of human resources, although the rates paid to those resources are not likely to be within the control of the project manager. In order to get the best return from the investment in those resources, the utilization of the resources assigned to the project team need to be seriously considered. In order to do that, the following actions must be taken:

- Identify strengths and weaknesses of human resources
- Provide training if possible that enhances their abilities
- Make good work assignments that are appropriate to the individual
- Monitor individual and team behavior
- Manage contract staff as well as public-sector employees
- Take corrective actions, and engage in progressive discipline if necessary
- Provide incentives that meet the needs of team members, within public-sector constraints
- Inspire team and individual performance

Human resource planning and management is, of course, dependent on the circumstances of the organization and the project. In most public-sector organizations, centralized human resource management functions are well-developed and dependent on highly documented processes, including the development of position descriptions, hiring practices and procedures, work rules, and, as mentioned earlier, disciplinary processes. In some public-sector organizations, union rules also apply to some workers, who belong to bargaining units. As a result, a substantial portion of human resource planning and management in a public-sector project is dedicated to interfacing with those existing systems and processes. The existence of those highly developed human resource systems is both an advantage for project managers, who are not required to develop their own job descriptions and other documents, and a challenge, because of the constraints those systems create.

In order to take an orderly and disciplined approach to the management of the project team, a human resources management plan can be created. That plan could include:

- The methods to be used to acquire the project team, if they are not already assigned
- The times in which resources are necessary, sometimes illustrated by the use of a resource histogram, a bar chart of resource utilization by time period
- Training needs and plans
- Team evaluation criteria
- Team rules
- The project organization chart

That plan should be as detailed as necessary to meet the needs of the project and clarify assignments and expectations. After creating that plan, the next step would be to acquire the project team. Acquiring the project team is highly dependent on whether team members have been preassigned and what options are available for selecting the project team. Very few public-sector organizations are organized to support projects, and most are functionally organized. That means that the selection of project team members is going to be highly influenced by functional managers. As a result, the project manager may need to negotiate to make certain that necessary resources are provided. In addition, many public-sector projects attempt to make use of existing resources that do not require additional, out-of-pocket costs. Those in-house resources may not have the optimal skill set.

In some cases, projects can be substantially handicapped by the lack of availability of key resources, because functional managers do not want to share them with the project. In most cases, as well, project resources assigned from other functional units will still be held responsible for their regular jobs, which will constantly draw them away from participation in the project.

A project manager can attempt to push the responsibility for acquiring an adequate project team up the line to senior managers by creating an assumption early in the project that adequate resources will be made available. Although senior managers can surely help make the case with other functional managers that resources must be made available, the project manager is responsible for project performance whether other managers deliver on their commitments or not. The project manager should always consult with human resource professionals in the agency to identify the required processes and the latitude allowed the project manager to counsel or discipline staff. Although human resource staff is sometimes the bearers of bad news, they are an important asset for the project team.

Motivating and Managing the Project Team

Given the types of constraints already identified for the management of human resources in public-sector projects, it should come as no surprise that motivating and managing the project team is a clear challenge. Once again, those challenges are not created by team members who lack motivation, but by the constraints on human resource management in the public sector and the many other competing demands on the time of team members.

As indicated earlier, the tools available to public-sector project managers for developing the team are sharply limited. Training budgets are typically limited, and training may not have been included in the project budget. In addition, the provision of training may be limited by departmental or higher constraints. For example, training may be centrally controlled and purchased, and training options may be limited to those programs approved by higher-level agencies.

Of greater concern, however, is the limited ability to improve feelings of trust among team members and create project buy-in. Many public-sector employees have been on the job for considerable amounts of time, and many of those long-term employees have developed a healthy skepticism about any initiatives. They have likely seen a lot of projects come and go and have seen many fail. As a result, some may approach project assignments with the expectation that this project, like so many others they have seen, will fail. Because of rigid compensation systems, most know that the ability of the project manager to reward them for exceptional service is limited or nonexistent.

Gaining the trust of team members in public-sector projects is a daunting challenge. Building that trust takes personal integrity (i.e., making commitments and living up to them) and the use of management techniques that speak to the needs of public-sector employees. As in any type of project, public-sector project managers should be aware that excitement about a project and full commitment to it may not be possible with some public-sector employees, but most public-sector employees will do their jobs in a professional manner.

Much of the work of public-sector projects can be defined as "knowledge work," which involves the manipulation of information and the creation of knowledge. Those who engage in knowledge work, which includes most public-sector employees and consultants, can be motivated by nonfinancial factors. Those workers enjoy learning, the creation of new ideas and concepts, and solving difficult problems; they are motivated more by the type of work than by the compensation offered for the work. Knowledge workers are also motivated by giving them control over their work and the opportunity to be creative. If public-sector project managers can insulate project team members from public-sector control systems that can interfere with effective work, they can allow project team members to have that level of control and creativity as a means of motivating them.

The types of intellectual challenges that public-sector project managers can offer to team members are actually motivating factors for knowledge workers. As a result, although opportunities for exciting the team with the potential for financial rewards are limited, many tough problems and interesting challenges can excite public-sector knowledge workers. In addition, because public-sector employees are motivated by the desire to serve the public interest, the project manager can help motivate them by linking the project to important public policy outcomes and the ability of the project to make a difference in the lives of citizens.

Managing the project team includes the activities necessary for supervising the work of the team, evaluating their performance, making shifts in staffing and assignments, managing contracts, and communicating with the project team. In managing team performance, the project manager is engaged in managing conflict, resolving issues, and inducing team members to perform as required. Conflict management and management strategies are discussed more in the next section.

Managing team performance also includes tracking and resolving issues. Issues are different from project risks or assumptions. Issues include decisions that have not been made and conflict that is yet to be resolved. Issues can be tracked by using an issue log. An issue log lists issues, persons responsible for resolving them, the actions they are to take, and dates by which the issue is to be resolved. An issue log allows the team to keep track of issues and their resolution. A disciplined approach to issue resolution allows the team to develop better methods for issue resolution, rather than relying on ad hoc approaches as individual issues arise.

An issue in a public-sector project may include something like "a functional manager has asked the deputy director to intervene to remove a critical resource from the project and reassign her back to the functional manager on a full-time basis because the team member's real job is not getting done in her absence." The issue log could identify the issue, identify what actions will be taken to resolve the issue, and designate the person responsible and the date by which it is to be resolved. That issue and the success in resolving it can be reviewed at a later date. A final output of managing the project team should be lessons learned, which could help manage future projects.

Resolving Project Conflict

You might assume that the best projects are those without conflict and that managing a project is simply a matter of getting people to perform as you direct them to. That viewpoint follows what is known as the old view of conflict. In that view:

- Conflict was always regarded as bad.
- It was deemed to be caused by troublemakers.
- The best managers were those who were able to create a smooth operation with little friction.
- People were viewed as the cause of conflict.
- Our tools for dealing with conflict were limited to forcing people to do things our way.

Although that viewpoint may have been useful when organizations functioned in the old command-and-control, static, mechanistic, tightly engineered organization envisioned by people like Frederick Taylor, it does not work today. That viewpoint assumed that people are interchangeable in an organization or process and that they could not add intelligent life to the organization or process.

Projects are full of conflict—among stakeholders, among team members, and even among the elements of the triple-constraint model. Now it is understood that, not only is conflict in project unavoidable, but conflict can be very healthy. It can help identify where issues are and uncover better ways to get things done. Today, we need a set of tools that allow us to get to the root of conflict and find options that create win-win solutions. We also need tools that can allow us to draw on the diverse viewpoints of our project teams without letting that diversity of opinion get in the way of getting work done.

In order to effectively manage conflict, we need to focus on the interests behind people's expressed positions. When people do not enter into a conflict they do not express what they really want. They usually begin by asserting a position, which is not what they want but, instead, is an argument they think they can win with. For example, if the human resources (HR) department is not getting what they want from a vendor and want to escalate the issue, they may demand that the contract with the vendor be terminated. That is probably not what they really want; they want the services they think they have contracted for. Instead of asking for what they want, they have taken a position that they think can get someone's attention.

In addition, our positions will be pumped full of emotion and exaggeration, which makes the conflict worse. Positions are usually backward looking, in that they focus on what has happened rather than what needs to happen in the future. Because our positions are often exaggerated, they are inherently dishonest and erode relationships. The only tool for dealing with position-based conflict is force.

For example, think about the HR example mentioned. The HR department has demanded termination of the contract, even though termination may not best serve their interests. By staking out that position, they have gotten the attention from senior managers they wanted and have probably gotten the attention of the vendor as well. That position, however, can induce other behaviors that make the conflict worse. The vendor, for example, might respond with its own position that demands additional payments from the agency. Senior managers might also react strongly by invoking contract termination clauses and alerting attorneys to prepare for legal action. The position taken by HR could, therefore, provoke responses and emotions that cause the conflict to grow out of control.

Fortunately, there is a better way. Behind every position is a set of interests that indicate what we really want or need. For example, the HR department may hold a set of interests that include:

- Being listened to
- Getting the services they need
- Being respected by the vendor

- Not being taken advantage of
- Reducing transaction costs
- Protecting the agency from legal challenges
- Making the project a success
- Complying with rules and laws
- Protecting personal reputations
- Avoiding being blamed for problems with the vendor

Notice that all of these interests are reasonable. Our problem with position-based negotiations is that people distill all of their legitimate interests into a dramatic position that might be unrelated to their real interests.

If we can get the parties in a conflict to express their interests rather than their positions, we can:

- Focus on solving problems rather than fixing people. People often take unrealistic and troublesome positions. When we analyze interests, however, we find that interests are almost always reasonable. That also helps take the emotion out of the conflict.
- Work on the real issues and interests rather than the smokescreen of positions.
- Broaden the solution set and try to find options for mutual gain. Our goal is not to make the conflict go away or even find a compromise where everyone gives up something. Our goal is to find a win-win solution that works to the advantage of both organizations.
- Create good relationships that can serve the project well in the future. To some extent, position-based conflict is based on childlike behaviors (a tantrum without a clear focus on real interests). Once parties have been exposed to principled negotiations, they may have the ability to solve conflict in the future more efficiently and with less emotion and pain.

In public-sector projects, this method can be particularly useful. Public-sector projects have a diverse array of stakeholders, who may have very different interests. Until those interests are identified, the problem being faced cannot be defined.

Too often, project managers make the mistake of trying to manage conflict by reacting to the positions that the parties have articulated. Better project managers take the time to step back from those positions to identify the interests of the parties. They find that the articulated positions rarely have much to do with the interests and the issues at the heart of the conflict.

STRATEGIES FOR MANAGING HUMAN RESOURCES IN PUBLIC-SECTOR PROJECTS

Fortunately for public-sector project managers, although the financial incentives for public-sector project team members are limited, several management theories have theorized that employees are not highly motivated by financial incentives anyway. These theories argue that employees are motivated more by the chance for self-actualization than they are motivated by salaries and promotions. Some of those theories are described as follows.

Using surveys of workers, Frederick Herzberg found that the factors that motivate employees are different from the factors that dissatisfy them. Workers, he found, were dissatisfied by factors, which he defined as hygiene factors, such as pay, working conditions, relationships with peers, and company policies. They were motivated by factors, which he called motivating factors, such as achievement, recognition, the work itself, advancement, and the opportunity for growth.

In the 1960s, Douglas McGregor framed a theory of management, which defined two disparate perceptions of how people approach work and how, as a result, they must be managed. The first, Theory X, views employees as seeking to avoid work. Theory X employees require supervision and direction, which those employees prefer over self-direction. Theory X employees value security over all else and can be motivated by the presence of penalties for nonperformance.

The other type of employees, which are defined as Theory Y employees, value self-direction and are self-motivated. Theory Y employees are capable of being creative in their attempt to solve organizational problems and want to accept responsibility. McGregor argued that better results could be obtained by employing strategies that assume that employees are predominantly from Theory Y. Importantly for public-sector projects, those Theory Y employees can be motivated by the chance to do interesting and important work.

McGregor's work was based, in part, on the work of Abraham Maslow. Maslow created the well-known hierarchy of needs. At the lower levels of the hierarchy are physiological needs, safety needs, and social needs. At higher levels are esteem and self-actualization. When a need is met, it ceases to be a motivator for employees. As a result, the best motivators are those higher on the hierarchy, although lower-level needs must be met before higher-level needs can have a motivational affect.

Victor Vroom defined expectancy theory, which says that, although employees have different goals, their behavior is a result of the extent to which they see a correlation between their own behavior and achievement of those goals. Vroom argues that employee behavior is a function of the extent to which they see effort tied to rewards, which, in turn, contribute to personal goals. Those personal goals are different for each person. For example, a person beginning a public-service career may be interested in being assigned to very challenging and demanding projects so that new skills can be acquired. A more senior person who is just a few years from retirement may not have the same interest.

The challenges for managers employing expectancy theory are to define the personal goals of employees, build trust so that the employees can see that performance will, in fact, lead to rewards, and identify necessary performance.

These theories of management provide hope for public-sector project managers in that they break the direct link between financial compensation and performance. They also emphasize the importance of interesting work and the chance for employees to make a difference. These are factors that public-sector projects often exhibit. Employees in public-sector organizations can often expect to be engaged in important work and have the opportunity to grow. In many cases, public-sector employees are given much more responsibility at earlier dates in their tenure than private-sector employees. Although this responsibility can be frightening, it can also be exciting and a motivator.

These management theories create challenges for public-sector project managers, however, in that they require that the project manager find a way to build trust with employees and create project buy-in. Those methods also require that managers foster the belief that employee performance will result in the rewards that employees value. Most public-sector managers are well aware of the deeply held levels of cynicism that many public-sector employees have learned over time.

PUBLIC-SECTOR LEADERSHIP

What is the difference between a project manager and a project leader? Although definitions of leadership and management sometimes blur, some common themes make them distinct. Leaders are regarded as risk takers, who seek opportunities, although the opportunities might not be well-framed. They focus on motivating people and framing the right questions for the organization. Leaders exercise both formal and informal sources of authority. Managers are focused more on solving identified problems. They focus on task accomplishment and the reduction of organizational risk. While leaders focus on doing the right things for the organization, managers focus on doing things right.

So which is the predominant role of the project manager in a public-sector project? For the most part, project managers, as the title indicates, are focused on those functions traditionally regarded as management functions. They are focused on solving the problem they have been assigned, with the least resources in the quickest time. Their goals are to accomplish the task set before them and to reduce the risk that the task will not be accomplished.

Project managers also have two leadership roles. First, most project managers have some say in the selection of projects and framing them for resolution. That requires that they identify opportunities and ask the challenging questions that can help the organization build strategies and develop projects that contribute to strategic outcomes. Second, project managers lead their project teams. Those project teams become organizations in their own right and require all of the leadership functions that are exercised at higher levels of organizations and agencies. Project managers are required to motivate and excite team members, build a vision for the project, and enhance the capacities of their team members.

Some project managers are better at one of these functions (management and leadership) than the other. Effective project managers have the ability to do both: to focus on the tasks at hand while leading their teams to do exceptional things.

BEST PRACTICES FOR HUMAN RESOURCE MANAGEMENT IN PUBLIC-SECTOR PROJECTS

Best practices for human resource management in public-sector projects include the following:

- Develop a human resources management plan
- Regard the human resource department as an asset rather than an impediment, and build an effective relationship with it

- Be demanding but respectful in getting the resources you need
- Assume that the necessary resources for the project will be available, document that assumption in the project charter, and make that assumption visible in status reports
- Give the project team as much latitude as possible in defining their own activities
- Engage project team members in project planning and project decision making
- If necessary, make a real commitment to engaging in the progressive discipline process, with the intention of improving performance
- Assume that your team members are Theory Y employees, unless given solid reasons to believe otherwise
- Give team members the chance to experiment, learn, and grow if possible
- Remember that the project manager has the ultimate responsibility for project success, but that success cannot be obtained without the support of the team
- Build a diverse team and respect that diversity
- Make good on your commitments as a means of building confidence that the work of the team will be rewarded
- Engage team members by reminding them of the importance of the project
- Protect the team from outside interference if possible
- Fix problems, not people.
- Never try to resolve conflict on the basis of positions
- Remember that emotions are also a factor in resolving conflict
- Identify the factors for creating high-performing project teams, and determine which of those factors can be applied

DISCUSSION QUESTIONS

1. What challenges have you encountered or can you imagine in the management of human resources in public-sector projects?
2. What challenges have you had in acquiring a project team with the right skills and the time necessary to work on your projects?

(*continued*)

3. What challenges have you faced in building a team orientation in public-sector project teams? What has worked and what has not worked?
4. What experiences have you had in the motivation of project team members in public-sector projects? Have you applied these theories or others? Under what circumstances might they work or not work?
5. What challenges have you faced in managing conflict in your projects? What means have you used to resolve the conflict? What elements of the principled negotiations method do you think would be useful for your projects?
6. What high-performing project teams have you been a part of? Which of these characteristics describe your experience? What other factors do you think could contribute to the creation of high-performing project teams?
7. What differences do you see between management and leadership? How do you think both apply to the role of the project manager in public-sector projects?
8. What other best practices can you imagine for managing human resources in public-sector projects?

EXERCISES

1. Create the necessary elements of a human resource plan for a project you are familiar with.
2. For the following situation, identify the parties, their positions, their interests, and some win-win solutions.

   Karen is the project manager responsible for managing the development of a new payroll system for the department. The project is relying heavily on a software vendor. Ed, the IT manager for the department, just came into her office.

   "That's it!" he shouted. "The vendor has missed another deadline on the software enhancements we need to create the new payroll system. Those people in payroll are about to rip my head off, and if

we don't get the system done on time, they claim that they might miss payroll. Can you imagine what a mess that will be?

"Karen, you've got two choices," he continued. "You can either terminate that contract or take the responsibility for missing payroll." He stomped out feeling that he had accomplished his mission.

Karen called Setha, the contract liaison for the vendor. "What's going on with the software enhancements for the payroll system?"

"Interesting that you called," said Setha. "I was just about to call you.

"To make those enhancements, we need several software modules from Ed's staff," she continued. "They provided us those modules on the day they were due, but I've never seen such junk. My kids could program better than that.

"I've had to put two additional programmers on the project to clean up what they gave me," she said. "I need to invoke the contract clause that allows us extra compensation for providing services not detailed in the service level agreement."

Rebuilding Greensburg "Green"

On the night of May 4, 2007, an EF5 tornado more than two miles across leveled the town of Greensburg, Kansas. When the sun came up, 12 people were dead, more than 95 percent of the homes and businesses were demolished, and the remainder of the buildings were uninhabitable.

After the rubble was cleared, and as Greensburg contemplated rebuilding, it had a completely blank slate, and it seemed natural that it should rebuild using emerging "green" building processes. The Greensburg city council approved a rebuilding plan that committed to LEED Platinum-level standards, making it the first city in the United States to commit to those standards of green building. Their goal was to create a sustainable community that was as energy efficient as possible.

A host of allies came to Greensburg's aid, including Governor Kathleen Sibelius and the Kansas Chapter of the American Institute of Architects. Plans were made to rebuild municipal buildings, schools, commercial buildings, and homes. In order to make sure

(*continued*)

that Greensburg had a sustainable economy, a business incubator was included in plans.

But Greensburg faced significant obstacles. The first was cost. Even in a declared disaster area, insurance and FEMA limit rebuilding costs to the replacement value of the home. That replacement value does not include conformance to green standards that can cost 3 to 5 percent more than normal construction. Although Greensburg had the advantage of rebuilding many homes, which could reduce the cost of construction, there is no proven model for building a community to meet green standards, and there are not even clear definitions of what green building means.

Part of the solution to the problems related to the costs of rebuilding green is a fundraising effort and website. The Greensburg rebuilding fund and its website (http://bigwell.org) provide information about rebuilding plans and opportunities for providing help. Another site called Greensburg Greentown (www.greensburggreentown.org) provides information and resources to residents as they attempt to rebuild, and the official Greensburg website also provides assistance and guides to city services, including building permits, codes, and registered contractors, as residents attempt to rebuild (www.greensburgks.org/resident/rebuilding). Corporations also stepped in to fund the gap between the available funds and the cost of green development.

Although the rebuilding process is not complete, the process is moving forward, and key building projects are being completed. The Greensburg rebuilding project demonstrates the value of establishing challenging but reachable goals, mobilizing the community and external stakeholders, and utilizing creative tools to accomplish project goals and public purposes.

Chapter 10

Managing Project Communications

According to some sources, communicating with stakeholders, which includes team members, takes up 90 percent of the project manager's time. Whether or not this is true, project communication is a critical function that can make or break a project.

It is also important to note that the modern model of project management requires project managers to (1) be aggressive in their communications so that they can keep stakeholders informed, (2) engage their team members so that their best talents can be optimized for the project, and (3) manage the expectations of a wide array of stakeholders so that everyone involved in the project understands what deliverables it will create and which deliverables are out of project scope. As noted before, the stakeholders of any project include those involved in the project, those who are affected by it, and those who are able to influence the outcome of the project.

In public-sector projects, stakeholders should be defined very broadly. The project team and others in the agency may be involved in the project. The users of the product of the project can include other agency staff, managers and staff from other agencies, the public, and future generations. Those who can influence the project include regulatory and oversight agencies, legislators, the public (again), and the press. As communication is given about the project, all of these stakeholders have to be satisfied, and an entire community with disparate interests must be managed in order to create project success.

THE CHALLENGES OF PROJECT COMMUNICATIONS IN PUBLIC-SECTOR PROJECTS

Communicating with stakeholders is a challenge for both private-sector and public-sector project managers. In both cases, stakeholders can have a wide array of interests. For example, some stakeholders may be interested in:

- How the project will affect their personal interests
- The protection of the status quo
- The maintenance of their personal authority
- Ensuring that key processes remain in place so that those processes can continue to meet their goals
- Decreasing their workload
- Increasing the workload of others

Although those interests may not be optimal for the purposes of the project, the project team will have to recognize that they may not be unreasonable interests from the perspective of the person holding them. (As we noted in the last chapter, interests are always reasonable to the person holding that interest. If we were in their situation, we might very well have the same set of interests.) Working with stakeholders who may hold interests such as these is one of the challenges of project communications management.

In public-sector projects, the challenges of project communications may be even more significant. In public-sector projects, the array of stakeholders may be broader and include legislators and the public, and there are likely to be more stakeholders who can influence the outcome of the project than there are in private-sector projects. For example, in a public-sector project designed to create something as simple as a tracking system for dogs in the county animal shelter, the stakeholders might include the following persons with their associated interests, as shown in Table 10.1.

These stakeholders range from internal stakeholders, much like those encountered in private-sector projects, to external stakeholders, including elected officials and the media. The key difference here is that for public-sector projects, more stakeholders are interested in finding instances of failure than for private-sector projects. In private-sector projects, internal politics may be encountered, but all of a project's stakeholders are likely to be interested in its success and the betterment of the organization.

Table 10.1 Stakeholders for Dog-Tracking System

Stakeholder	Interest
County commissioners (elected)	Constituent satisfaction, budget impact
County administrators	Budget impact, utilization of resources, sustainability of the solution
Agency staff	Impact on workload, ease of use of the system
The press	Potential for a good story (e.g., project failure or wasted funds)
Dog lovers and activists	Impact on dogs, incidents of impact on dogs that can be used to leverage their position
County purchasing administrators	Compliance with rules and procedures
County legal counsel	Compliance with rules and laws, reduced liability for the county
Opposition party	Opportunity to find information that can be used against current office-holders in the next campaign
Cat lovers	Equity for cats
Vendors	Opportunity to earn a profit
Project team members	Interesting work, project success, creation of realistic expectations
Project manager	Success of the project

As a result, public-sector project managers must:

- Overcommunicate with many stakeholders
- Respond to the varied interests of the stakeholders with different communication styles and methods
- Exercise caution in communications so as to avoid political exposure or embarrassment
- Manage communications with vendors to ensure compliance with legal requirements and processes
- Identify accessible points of contact and those who can assist in communications with stakeholders, such as aides to elected officials

Although a project manager may not have access to all of the critical stakeholders and expertise in managing each one of the stakeholders, project managers for public-sector projects need to identify the interests of all of those assorted stakeholders and find allies who can help them interact with them.

Public-sector projects are also confronted with two contradictory requirements. Some projects require high levels of secrecy. Those sensitive projects could include military, law enforcement, and security operations. Other projects are required to make information available to the public when requested. In the United States, Freedom of Information Act (FOIA) requests can be filed by anyone. In response to an FOIA request, the project team may need to provide the requested information.

THE FUNCTIONS REQUIRED FOR PUBLIC-SECTOR PROJECT COMMUNICATIONS MANAGEMENT

Three required functions were identified for the successful management of project communications for public-sector projects. They were:

- Creating a communications plan
- Distributing information to stakeholders
- Capturing and managing knowledge

Each of these functions is described in turn and their implications for public-sector project management examined.

Creating a Communications Plan

Although communications planning is required for all types of projects, it is even more essential for public-sector projects. As indicated earlier, the communications networks and overlapping levels of oversight applied to public-sector projects ratchet up significantly the need for and difficulty of communicating effectively with all of the project stakeholders. Creating a communications plan involves the identification of stakeholders, determination of their needs, and development of plans for meeting those needs. The ultimate output is the creation of a communications plan that can be used to guide project communications and track project performance.

The communications plan for some public-sector projects will not be highly formal, although, in some cases, a formal document will be required. Whether the plan is formal or not, the mental exercises required to create the communications plan are required of all projects. Every project needs to consider who its stakeholders are, what their interests are, and how they can best be communicated with, even if a written document is not created.

Project managers for public-sector projects need to understand that stakeholders vary in their need for information and the way that they prefer to be communicated with. Simply firing out e-mails containing project status reports will not likely be an adequate strategy for all stakeholders. Especially in a political environment, stakeholders will require personal communications and messages tailored to their needs. In a public-sector project, the project manager should expect to spend a considerable amount of time communicating with stakeholders and should build the necessary time for that communication into the project plan.

The project communications plan can consist of:

- Identification of the stakeholders
- Their role in the project
- Each stakeholder's interests in the project (e.g., fiscal staff may be interested in the conformance of the project with budget limits and adherence of the project to financial rules, while key legislators may have an interest in the impact of the project on their constituents)
- Information the stakeholder will require (e.g., a detailed status report, a special report on constituent issues and feedback, description of the product of the project, a press briefing, or a short summary of project status)
- How often that information is needed (e.g., monthly, weekly, daily)
- How that information will be provided to the stakeholder (e.g., via e-mail, a regular briefing, a discussion with legislative aides, or a project newsletter)
- Who will provide that information

The appropriate use of information technology is an issue that project managers need to carefully consider. In general, although electronic communications technologies are useful, project managers should not overlook the value of personal communications, even though they may be time consuming. Some stakeholder groups in public-sector projects are notably recalcitrant in their willingness to use electronic means of communications. That is particularly the case in some citizen groups, who may be in most need of services. In addition, public-sector project managers may need to consider the use of multiple languages in their communication with citizen groups.

Table 10.2 Communications Planning Template

Stakeholder	Interest in the Project	Information Requirements	Frequency	Method of Information Provision	Person Responsible

Electronic data repositories and means for document sharing and collaboration are becoming increasingly sophisticated and useful for project teams. Having a repository that is web-accessible can be very useful for storing and sharing project documents, communicating with team members, collecting useful information, and allowing collaboration on documents and the sharing of ideas (see Table 10.2). Many technologies are available, and most can be assembled by the project team without extensive assistance from technology experts. More will be said later about the need to capture project information and share it for the benefit of later projects.

Providing Information to Project Stakeholders

Information distribution implements the communications plan and responds to communications requirements that are not included in the plan. For example, new stakeholders may be identified after the plan is developed and new issues may arise that require extensive communications efforts. It is not uncommon for new parties with an interest in the project outcome to reveal themselves well after the project is initiated or for new regulatory requirements to be identified. The project team has several options for choosing the best information distribution methods.

In large, public-sector projects, project managers need to be especially creative in finding ways to communicate with large numbers of stakeholders. The use of periodic press conferences, public forums for gathering input, and newsletters is not out of the question, although the use of those mechanisms should be built into the project plan and approved by senior managers. Even electronic means of communications can be used creatively. An excellent example is the Commonwealth of Virginia's

Department of Transportation "project dashboard."[1] That dashboard provides a wide array of highway project information in a colorful and graphic-intensive format.

Project managers also need to consider the formality or informality of their communications. In general, two factors lead us to more formal mechanisms for communication: the level of risk involved in the communication and the level of noise that could disrupt the communications.

Some project communications are of relatively low risk, whereas others have much higher levels of risk associated with them. For example, where there is a risk that legal action might be contemplated or is actually being engaged in, communication risk is very high. That is, we want to be very clear about what we have said. In those cases, communications are likely to become more formal.

Overall, because of the detailed and structured processes required for public-sector activities and the requirements for compliance with those processes, communications in public-sector projects are generally more formal than those in private-sector projects. That extra level of formality is both an advantage (e.g., the creation of a solid written record about the project) and a detriment to the project (e.g., the extra time required and sometimes unduly complicated communications).

Noise is anything that can disrupt the communications channel. Noise in project communications can be literal, which is the case in military operations, or figurative. Figurative noise that can disrupt communications includes such factors as language differences, lack of understanding, different technical aptitudes, differential understanding of processes and issues, and different political perspectives. Several strategies exist for managing noise.

The most prevalent strategy for reducing the impact of noise is redundant communications. Redundant communications require that both the sender of the message and the recipient of it assume two responsibilities. The sender must (1) send the message in terminology that the recipient can understand *and* (2) ensure that the message was received correctly. The recipient must (1) receive the message *and* (2) ensure that it was received correctly. Redundant communications requires that the sender and the recipient engage in a dialogue that extends beyond the simple transmission of the message.

---

[1] The dashboard is available at http://dashboard.virginiadot.org.

For example, if the project manager wanted to remind a team member that a key deliverable is due next week, he or she could send an e-mail that simply says "Hey, I just wanted to remind you that the deliverable is due next Tuesday." The team member might receive the e-mail, make a mental note, and delete the e-mail. Note that there is no assurance that the message was received and that any action will be taken. Alternatively, the project manager could add a note to the e-mail that said, "Let me know today how you're doing and what needs to be done to get the deliverable done by Tuesday." That additional request requires the recipient to respond and ensures the project manager that the team member has received the message and is aware of the requirement.

If only the first message were received by the team member, and the team member wanted to engage in redundant communications, the team member could send an e-mail back to the project manager saying, "I assume you mean deliverable 4.2 and that you want me to deliver it directly to the deputy director." By making that response, the team member has clarified the intention of the project manager.

Without question, all of this dialogue takes time and requires, in many cases, that people get out of their comfort zones and engage in dialogue with others. The alternative is that we merely assume that the message was received and leave to chance the possibility that noise has overwhelmed our communications to the detriment of our projects. Note that the purpose of redundant communications is to help make the project successful. It is not a tool that is intended to fix blame or establish "gotcha" messages.

As we are distributing information about the project, it is also important that we take the time to disseminate lessons learned. Good project management facilitates organizational learning, which is critical in a fast-paced environment and in an environment in which high staff turnover is a risk. That organizational learning and the ability of the organization to increase its project management maturity depend on the commitment of every project to reflect on what it has learned, both about what has worked and what has not worked. Lessons learned can be accumulated throughout the course of the project and need not wait until the end of the project to be identified and documented. If lessons learned can be identified during the course of a project, they might actually be used to improve the management of that project.

Performance reporting is one specific type of project communications that involves collecting data on the performance of the project and reporting that data to project stakeholders. That performance data should include schedule, cost, scope, and quality data. In most cases, status reports will be prepared on a regular schedule, although special-purpose status reports can be necessary. In public-sector projects, compliance reports will also be necessary to show the extent to which the project has met legal and other requirements, such as compliance with minority set-asides for project procurement.

Status reports can also be extended to include predictions of future project performance based on information collected thus far. The provision of a status report provides a good opportunity for reviewing performance and assessing future trends and likely project costs and end-dates.

The selection of information to include in status reports is a challenge. Too often, project managers adopt the default option and report everything that is known about the project. Those reports can run to multiple pages of small print, which overwhelm stakeholders and cause some to disregard status reports completely. The best project status reports provide only the necessary information in a format that is visually appealing and easy to read. Some elements that might be included in a short project status report include:

- A short project narrative
- Schedule milestones
- Deliverables completed and pending
- Critical risks
- Critical project assumptions
- Budget status

Multiple status reports may be necessary to meet the needs of various project audiences. One organization that has reached a high level of project management maturity sets two rules for its project status reports. Those reports:

- Should be as visual as possible and minimize text
- Must be limited to one page

Public-sector project managers also need to be aware of the key role that aides and assistants play for busy managers and legislators. Often the best (and sometimes only) way to communicate with a senior stakeholder is to brief their aides or assistants. Although aides or assistants may seem to be young and relatively inexperienced, they play a key role in identifying important issues and filtering information for the senior person they work for. Building a good relationship with an aide or an assistant can pay dividends for a project.

As project managers distribute information to their stakeholders, they need to keep in mind the need to manage the expectations of those stakeholders. It is sometimes tempting to provide stakeholders with glowing reports of project accomplishments. Those glowing reports can, however, come back to haunt the project manager if things do not turn out so well later.

Stakeholders need accurate but understandable information about the project. Reports that are overly optimistic can cause stakeholders to develop their own aggressive expectations and could cause resource reductions if stakeholders feel that there are more pressing needs for the resources in other projects. If things are going well, we should report our progress but identify the realistic challenges we still face. Reports that are overly pessimistic can cause a loss of stakeholder support and skepticism about the project. If things are going poorly, we need to be honest and identify our plans for correcting problems. If those plans require action on the part of stakeholders, those required actions must be clearly identified and requested.

Because of the diversity of stakeholders for public-sector projects, managers of those projects probably require higher levels of these communications skills than other project managers. Those skills are difficult to teach but priceless if they can be mastered. Managing stakeholder expectations also requires that the project manager have a good sense of the organization and the ability to navigate through the intricacies of the hierarchy. That is to say that one of the most valuable skills for any project manager is the ability to get things done in the organization. That skill requires an understanding of the goals of the organization, its processes and rules, the key decision makers and change agents, and the critical organizational "minefields" that can disrupt the project.

Too many inexperienced project managers operate under the "empty bucket" assumption. That assumption implies that people have a figurative empty bucket on top of their heads and that all we need to do to communicate with them is to toss our message into their buckets. Project managers who simply e-mail project data to stakeholders and expect them to glean necessary information from that data operate under that assumption. Of course, no one has an empty bucket. Our buckets, instead, are overflowing and, for someone to get a message in our bucket, they have to compete for space in the bucket. When they simply toss a message our way, it will either fail to get in the bucket or, even if it gets into the mix in the bucket, it will be diluted by the other messages already in there.

Our challenge in today's high-speed, high-stress public-sector workplace is to compete successfully for the time and attention of stakeholders for our projects. Getting that attention requires that project managers find creative ways of communicating and to invest the time in communications that is necessary to keep stakeholders aware, satisfy their information needs, and create broad project understanding.

Capturing and Managing Knowledge

Early in this book, some of the challenges for government that are looming in the near future were identified. In addition to the external challenges being thrust on government agencies, which include reduced revenues and increasing demands for services, there is an internal challenge as well that could, if not addressed, diminish the ability of public-sector agencies to deliver services.

That challenge is the loss of organizational knowledge, caused, in part, by the impending departure of significant numbers of the members of the baby boom generation. In addition, the knowledge necessary for successful agency operations is becoming more difficult to capture as vendors are used for more critical projects and the useful life of knowledge becomes shorter.

All sectors of the economy are affected by this loss of expertise and organizational knowledge, but the public sector is impacted particularly hard. The U.S. Office of Personnel Management, for example, predicts that 37.3 percent of the 2006 Federal workforce will retire by 2016. That number is only 61.3 percent of those eligible to retire by that date, which could drive the count higher.[2] For project managers, these retirements and other departures from the public service increase project risk. Although project cost risk could decrease as high-cost employees retire, other risks go up. Those risks include the risks that:

[2] United States Office of Personnel Management, "An Analysis of Federal Employee Retirement Data: Predicting Future Retirements and Examining Factors Relevant to Retiring from the Federal Service," March 2008, p. 6.

- Quality standards will not be met.
- The project team will overlook rules and processes with which the team must comply.
- Human networks that facilitated project work have been diminished.
- Project costs will increase as team members learn on the job and are less efficient in producing outcomes.
- The project will be delayed because of a decline in worker productivity.
- Lessons learned in prior projects will not be accessed because of the loss of organizational memory.

Many public-sector agencies possess little in the way of capital assets, and, as a result, the most significant asset they hold and that contributes to the performance of agency work is the combined knowledge of the workforce. A similar argument can be made for public-sector project teams. In order to optimize project outcomes and protect the agency, public-sector project managers have to give special attention to the management of knowledge as it relates to the project.

Two types of knowledge can be lost in organizations—explicit and tacit knowledge. They require different strategies for retaining and reusing knowledge.

Explicit knowledge is knowledge that can be written down. In public-sector projects, it can include knowledge of laws, rules, processes and procedures, agency products and services, and job descriptions and classifications. Strategies for project managers attempting to capture and reuse explicit knowledge can include:

- Creating a shared, electronic space for the project team and stakeholders that allows them to access project documents and share their own observations and experiences
- Creating standard project information and reusable templates that can be filed at the close of the project and used by future project teams
- Creating project management methods and templates that can be documented, updated, and reused
- Taking seriously the processes of sharing knowledge and documenting lessons learned

Tacit knowledge is harder to capture. Tacit knowledge is the knowledge that we may not even know that we possess. It includes the knowledge of context and background. Tacit knowledge is more personal and must be transmitted through personal contact between those who hold tacit knowledge and those who need it.

In public-sector projects, tacit knowledge allows individuals to work together and to get work done. It can take the form of networks of relationships held together by knowledge and deep background on the context of the project. Often, those with high levels of tacit knowledge provide guidance on the informal networks and patterns of decision making. Tacit knowledge also allows the project team to avoid the organizational minefields that create high project risk.

The transfer of tacit knowledge requires high levels of personal interaction, which is why it is so difficult to identify and why it can be diminished by the departure of senior staff persons. Methods that a public-sector project manager might apply to capture, transfer, and enhance tacit knowledge include:

- Establishing mechanisms that allow less-experienced project team members to be mentored by senior team members
- Creating opportunities for professional and social interaction among multigenerational team members
- Seeking out those in the organization who are likely to have accumulated significant amounts of tacit knowledge and interviewing them with an eye toward helping them discover all that they know about the organization
- Taking advantage of professional networking opportunities outside of the organization for project team members, in that tacit knowledge is not exclusively related to the home organization
- Creating reward systems that encourage conversation and sharing of tacit knowledge and a project environment that allows for dialogue
- If possible, providing a physical space for brainstorming and sharing by team members

In short projects, the positive impact of knowledge management may not be noticeable, but in longer projects and in a succession of projects, knowledge management can pay significant dividends. As the impact of the loss of knowledge becomes clearer, public-sector agencies can be expected to enhance their tools for retaining and managing knowledge, which may include more formal mentoring and employee-retention programs.

BEST PRACTICES IN PUBLIC-SECTOR PROJECT COMMUNICATIONS MANAGEMENT

Best practices for managing communications in public-sector projects include the following:

- Identify the project's stakeholders and their varied interests
- Create a communications plan
- Enlist allies for communicating with senior-level stakeholders, including legislative affairs experts, if appropriate
- Create short project status report formats and carefully consider the most important information to present
- Create multiple reporting formats if appropriate but ensure consistency among them
- Recognize the important role of aides, assistants, and secretaries in getting information to senior persons
- Employ redundant communications and proactively address noise in the communications channel
- Maintain an issue log
- Use more formal communications strategies for high-risk communications
- Use creative techniques for communicating with stakeholders
- Build knowledge management into the project plan
- Create strategies for capturing explicit and tacit knowledge

DISCUSSION QUESTIONS

1. Who are the stakeholders of your projects? Is the public involved? What about oversight agencies, the press, and legislators?
2. What challenges have you faced in managing project stakeholders? How have you resolved problems? What skills did you apply?
3. What are the interests of your stakeholders? Were any of those interests surprising to you?
4. What noise disrupts your communications?
5. What means of communications have you used to reach the public and the media? How have you interacted with oversight agencies to keep them informed?

6. What strategies have you used for capturing and managing knowledge? What strategies and mechanisms do you employ for explicit knowledge? Tacit knowledge?
7. What other best practices for managing communications in public-sector projects can you imagine?

EXERCISES

1. For a project you are familiar with, create a project communications plan using the template provided. Pay special attention to the needs of stakeholders, like the press or legislators, who cannot be reached using standard communications methods.
2. For a project you are familiar with, create a one-page status report. Choose the elements of the report carefully and identify where charts or graphs might be used to communicate information.

The Manhattan Project

Even if you disagree with the goals of the Manhattan project or are dismayed by its outcomes, it provides useful lessons for the managers of public-sector projects and has to be admired for its ability to produce its intended outcomes under very tight timelines.

The Manhattan Project was initiated by the United States among fears that Nazi Germany was developing a nuclear weapon of its own, a fear driven by the published research of German scientists before the war. World War II erupted just as atomic energy progressed from being possible to being likely. In 1942, J. Robert Oppenheimer was asked to begin to coordinate research in the United States, and he convened a summer conference at the University of California at Berkeley to identify the feasibility of a fission bomb. That conference determined that such a bomb could work.

As the project developed, the need for centralized control and transfer to the military became clear, because the project was intended to create a weapon, rather than scientific knowledge. President Roosevelt

(*continued*)

assigned the project to the Army Corps of Engineers, and Colonel James Marshall was put in charge. When Marshall failed to meet expectations and get the project moving, it was reassigned to Colonel (later General) Leslie Groves, who solved two of the project's most urgent problems in the first two weeks on the job.

Groves, however, was regarded by the scientists working on the project as pushy and overbearing. Only after the project was over did they begin to appreciate his leadership. In October 1942, Groves asked Oppenheimer to supervise the new laboratory for physics research and design of the weapon because of his ability to lead scientists. Both began to recognize the value that the other brought to the project.

Three critical and related deliverables were required. They included the development of the bomb, the production of nuclear material, and the planning for dropping the bomb, which included modifying aircraft, training crews, and securing a takeoff site so that the aircraft could reach its target.

The project goals were met on July 16, 1945, when a nuclear weapon was exploded in the desert near Alamogordo, New Mexico. It succeeded because of the significant support provided by the civilian leadership, the work of the scientific community, and the complementary management efforts of Groves and Oppenheimer.

Chapter 11

Managing Project Risk

To a large degree, this book, like most books about project management, is about reducing the risk of project failure. As noted earlier, project management methods are applied in order to reduce risk. Our basic concern as we consider establishing project management discipline is reducing the risk that our projects will fail, that they will fail to meet budget or schedule targets, or that they will fail to deliver products that meet the needs of customers.

Earlier, we discussed risk as the criteria by which we determine the degree of project management rigor to apply to our projects and argued for a risk-based analysis of project management methods. The goal in that exercise was to apply project management methods in direct relation to an assessment of the potential risk of the project. Those projects with high risk demand more rigorous application of project management methods. Those projects of relatively low risk can be managed less formally and with less strenuous application of project management methods.

This chapter examines methods for identifying project risks, analyzing them, and developing strategies for responding to those specific risks. The goal is to manage risks in a cost-effective manner in order to increase the chances of project success.

THE CHALLENGES OF MANAGING RISKS IN PUBLIC-SECTOR PROJECTS

Managing project risk requires that we spend some time envisioning those factors that could cause the project to fail or to be compromised. Once we have identified those factors, we need to create response plans to reduce their potential impact or probability. Our goal is to manage risk proactively rather than being passive to risk, which places us in the position of reacting to conditions that we might have been able to anticipate and plan for.

Managing project risk, those points of uncertainty that might impact a project, is always a challenge in either public-sector or private-sector projects. The final chapter examines ways to deal with high levels of uncertainty in public projects—uncertainty that borders on or creates project chaos.

Although both private-sector and public-sector projects have risks, public-sector projects may have additional risks that do not often occur in private-sector projects. Imagine that a state department of mental health and mental retardation were to attempt to maximize federal reimbursements for federally funded care for patients in state hospitals by building a system that collects direct and indirect costs, matches those costs to services provided to specific clients, and reports those costs in a way that meets federal reimbursement standards. Even for a similar private-sector project, substantial risks would apply and could include the risks that:

- Software would not be adequate for the project's requirements.
- New, costly hardware would be included.
- System interfaces would be overlooked or impossible to manage.
- Management would lose interest in the project.
- Cost overruns could occur.
- Systems might not comply with regulatory or statutory requirements.
- The data might not be available in the formats necessary.

In a public-sector application, however, other risks could include the risks that:

- The press might investigate the project and make public its challenges and failures.
- Necessary hardware and software might not be available through state purchasing systems.
- Salaries might not be adequate to attract expert staff.
- Federal policies and reimbursement models that the system was designed to meet might change.

- The project might be cut from the budget or might not be reauthorized if it crosses budget cycles.
- The best vendor might not be on approved vendor lists.
- Set-aside programs might require the use of less-qualified vendors.
- Opposition legislators might identify the project for electoral criticism.
- Less-qualified vendors might exert political pressure to gain business.
- Providers in the service-delivery network, such as county or private-sector providers of service, might not be able to comply with system requirements or might elect to develop their own systems or refuse to cooperate with this one.
- Systems might not be allowed to be created because they are incompatible with centrally mandated system architecture.
- Processes for hiring staff, contracting with consultants, and purchasing software and hardware might be slow and delay the project.

Those types of risks are particularly difficult to manage, because they extend outside the project and involve stakeholders with varied interests and motivations. In some cases, the project manager will be required to enlist allies and build a consortium of supporters who can influence others on his or her behalf. Managing these types of risks also requires a broader skill set than managing traditional project risks and an understanding of organizational processes and political systems and realities.

In public-sector projects, the points in the project likely to be the riskiest are those at which another department or agency is required to complete key processes. Examples are purchasing processes, hiring processes, legal processes, and budgeting processes. At these points, which can be very time-consuming, the project manager may feel relatively powerless. Because these processes cannot be directly controlled by the project team, the project manager will be required to rely on his or her ability to communicate with others, build effective relationships and allies, navigate those processes, and convince others of the need for timely performance.

It is important to note that risk management deals with "known, unknowns," those things that we can identify that we do not know. Dealing with "unknown, unknowns" requires managing circumstances when they later become apparent. The final chapter also addresses the management of those more challenging uncertainties.

THE REQUIRED FUNCTIONS FOR PUBLIC-SECTOR PROJECT RISK MANAGEMENT

Earlier, we identified five required functions for public-sector project risk management. They were:

- Creating a plan for risk management
- Identifying risks
- Analyzing risks
- Developing risk responses
- Creating a plan for managing the project's legal and administrative constraints

Each of these functions and their implications for public-sector projects will be discussed in turn.

Creating a Plan for Risk Management

Risk occupies a special place in project management. As noted earlier, nearly everything we do in project management is intended to reduce the risk of project failure. We create a project scope to reduce the risk of not delivering acceptable products and the risk of wasting effort on deliverables outside the scope of the project. We build a schedule to reduce the risk that the project is late, and we build a budget to reduce the risk of spending more than we are authorized to spend. We even build a communications plan and give attention to our communications with stakeholders to reduce the risk of failing to meet the information needs of our stakeholders.

Risk management is also a critical function, because it creates new activities that require resources and take time. As a result, risk management interfaces tightly with the other project management processes. Furthermore, risks are always changing, which requires us to constantly engage in risk management. Some risks identified early in a project may disappear later in the project. For example, an early project risk may be that legal approval will not be granted for contract documents. That particular risk should be removed from consideration once those documents have been approved.

Other risks, not identified early in the project, may become clear later. During early project planning, we might not have anticipated changes in federal law. If relevant law were to change, however, a host of new risks would emerge and would need to be reflected in project planning, including scope, time, and cost adjustments.

As a result, we have to adopt a proactive approach to risk and an approach that constantly evaluates and reevaluates our project risks. Risk management planning is not, therefore, a one-time effort, as might have been the case with quality planning or communications planning (though both of those plans might also require periodic updating during the course of the project).

Risk management planning does not deal with specific project risks. Instead, risk management planning creates the risk management plan. That plan for a public-sector organization can include items such as:

- Methods to be used to identify risks
- Categories of risks, including political, legal, and media risks
- Scales for ranking risks and identifying which risks are most critical
- Methods for managing the unique constraints of public systems
- Responsible persons for managing risks
- Risk terms
- The format of risk reporting systems
- Risk checklists created by prior projects

Some organizations will have developed risk management processes that are standard across all projects, which reduce the amount of time required for risk management planning. Risk management plans can be formal or informal but must induce consideration of how the project will manage its risks. Too often, not enough attention is paid to risk management planning, or it is assumed that everyone knows what the risks of the project are, which negates the need for thinking about risks and planning for them. Unfortunately, our automatic assumptions about risk are too often wrong.

For example, in information technology projects, team members will often identify the risks of inadequate resources, software incompatibility, missing interfaces, or lack of project prioritization. In actuality, however, the major reason for information technology project failure is the failure to meet the requirements of system users. Rarely do information technology project teams identify those risks and methods for reducing those risks by doing such things as ensuring that user requirements are identified, documented, and delivered. We can force them to think of requirements risks by including that category of risks in the risk management plan.

Identifying risk categories in the risk management plan helps broaden team member thinking in risk identification. Some risk categories for public-sector projects might include:

- *Schedule risks*: those factors that could delay the project
- *Cost risks*: those factors that could increase project cost
- *Scope risks*: those factors that could cause the scope to increase or to be inadequately identified
- *Political risks*: those factors that could cause political scrutiny or opposition
- *Publicity risks*: those factors that could cause the project to receive unfavorable press coverage or adversity
- *Stakeholder risks*: those factors that could cause stakeholders to lose support of the project
- *Process risks*: those factors embedded in purchasing or hiring processes that could delay or otherwise impact the project
- *Budget process risks*: those factors that are created by budget processes that could impact the project
- *Resource reduction risks*: the risk that project funding will be reduced prior to its completion, an increasingly common occurrence in public-sector projects

Stakeholder risk tolerances identify which risks stakeholders have more or less tolerance for. In public-sector projects, a few assumptions about stakeholder risk tolerances can be made. First, senior stakeholders have low tolerance for any risks that might contribute to project visibility, adverse press coverage, or political scrutiny. Second, public-sector project stakeholders are often not sensitive to cost risks because of the common practice of not identifying costs incurred internally. If external costs are involved, the tolerance for overruns may decline substantially, particularly if the project is budgeted as a line item.

Last, the tolerance for schedule risks may depend on whether externally imposed deadlines are included in statutes, rules, or higher-level government requirements. For example, federal Medicaid programs often require state-level compliance by a certain date. Failure to meet that compliance date can jeopardize the flow of funds.

Definitions of risk probability and impact, which can be defined in the risk management plan, help define parameters for ranking project risks.

Table 11.1 Risk Probability Scale

Probability Description	Probability Score
Nearly certain that the risk will occur	10
Highly likely that the risk will occur	7
Likely that the risk will occur	5
Not very likely that the risk will occur	3
Very unlikely that the risk will occur	1

Later we will identify project risks and discover that not all risks are of equal importance. Some are more important or serious than others.

Two factors determine the relative seriousness of a negative risk: (1) the probability that the risk will occur and (2) the impact it may have on the project if it does. If we can assign a probability to each risk and assign a numerical indicator of impact, we can multiply those factors to create a risk factor. Those risk factors can be compared to identify the relative ranking of the risk and prioritize the risks for the creation of risk responses. In addition, because we can identify more risks than we can find cost-effective responses for, we need to be able to prioritize risks to determine which we will act on.[1]

Risk probability is relatively easy to measure, because it is normally expressed as a numeric term (e.g., a 40 percent chance of the risk occurring). If organizations are not comfortable with defining percentages, they can create a simpler scale for converting the probability of risk occurrence to a numeric indicator. For example, they could use a scale like the one illustrated in Table 11.1.

Identifying the potential impact of a risk is more challenging, because impact is not usually described in quantitative terms. Instead, we are more used to describe it as "high" or "low."

We have a couple of choices for converting an estimate of impact from qualitative terms to quantitative terms. The first is to convert the entire impact into a financial impact. In theory, all impacts can be reduced to a

[1] Mathematicians quarrel with the multiplication of risk probability, which is usually by definition a quantitative measure, by the risk impact, which is a non-quantitative factor converted to a numerical indicator. Although they are correct in criticizing the multiplication of a quantitative term by a qualitative term, as long as we use these factors to compare risks within the project, we can still use the product of these two terms as a relative indicator of risk.

financial impact. For example, if we are exposed to the risk that a key project team member will leave before the conclusion of the project, we could estimate the costs of hiring a replacement and the costs of any rework that might be required.

In the private sector, many risks can be converted to their financial impact because of the explicit financial goals of those organizations and their propensity to deploy qualitative systems for evaluating projects. In public-sector projects, converting every potential risk to its dollar impact can be difficult. What, for example, is the dollar cost of disappointed citizens or lost credibility for the department? What is the positive impact of increased citizen participation in government?

A better way to identify the relative impact of the risks is to create a risk-impact table that allows a variety of factors to be included in the scoring of the risk impact. An example is illustrated in Table 11.2.

Table 11.2 Illustrative Negative Risk-Impact Scale

Criteria	Qualitative Value	Score
Project delay of over six months, budget impact of over $100,000, very significant impact on customer satisfaction; very significant political or media visibility; very significant potential to create noncompliance with law or administrative rule	Very high	10
Project delay of over three months, budget impact of over $50,000, significant impact on customer satisfaction; significant political or media visibility; high potential to create noncompliance with law or administrative rule	High	7
Project delay of over one month, budget impact of over $25,000, medium impact on customer satisfaction; medium political or media visibility; medium potential to create noncompliance with law or administrative rule	Medium	5
Little project delay, little budget impact, little impact on customer satisfaction; little political or media visibility; little potential to create noncompliance with law or administrative rule	Low	3
No project delay, no budget impact, no impact on customer satisfaction; no political or media visibility; no potential to create noncompliance with law or administrative rule	Very low	1

We could also create a similar scale for positive risks, which identifies potential positive outcomes, such as cost savings or reductions in the schedule.

Obviously, each organization can create a scale that includes risk evaluation criteria that are relevant to it and its stakeholders. In addition, the scores can be adjusted to indicate risk tolerances and can be either linear or geometric (e.g., the top score could be 100 and the bottom 1).

Later, we will multiply the probability and impact score for each identified risk to create a relative risk ranking for each risk. That risk ranking, or risk factor, will allow us to prioritize our project risks.

Identifying Risks

Risk identification allows us to create the risk list, that list of all of the risks of the project that we can identify at the present time. As a result, risk identification needs to be performed periodically throughout the project, as risks change and additional risks become identifiable. Risk identification should involve most of the project team so that team members can feel ownership of identified risks and so that the project manager can draw on the expertise of the team and its various perspectives.

Risk identification can be accomplished by using such strategies as:[2]

- Brainstorming
- Reviewing project documentation
- Analyzing root causes (Ishikawa diagrams)
- Analyzing checklists created for this project or prior projects
- Analyzing assumptions, which, as explained earlier, are related to risks in that they deal with uncertainty

Risk brainstorming can be more effective if it uses the risk categories identified in the risk management plan. Recall that we also need to identify positive risks, which are those factors that could lead to good project outcomes. Later, we will identify strategies for maximizing the outcomes related to positive risks.

---

[2] Project Management Institute, *A Guide to the Project Management Body of Knowledge (PMBOK® Guide)*—Fourth Edition, 2008, pp. 286–287.

The principal outcome of risk identification is the risk register. The risk register is the principal tool used to track project risks and is created in a stepwise manner. At this point in the application of the risk management processes, the risk register may only include the identified risks. Other elements will be added later. Using spreadsheet software to create the risk register allows the risks to be sorted from highest to lowest risk factor once those risk factors are calculated.

A risk register is illustrated in Table 11.3.

Analyzing Risks

Creating a thorough list of risks is a good idea, but what are we supposed to do with a list of risks that is probably too large for us to cope with? Our first step is to prioritize those risks. Not every risk is of equal importance, and to take a first cut at identifying the most serious risks, we need to do qualitative risk analysis.

Our principal strategy for qualitative risk analysis is the application of the probability and impact weighting described in the risk management plan. By gathering a group of stakeholders or team members together, assigning probability and impact scores to each risk, and multiplying those scores to determine the risk factor for each risk, a list of risks can be created that can be prioritized by the seriousness of the risk. We can also assign simpler indicators of risk seriousness by identifying those risks above a risk threshold as being "red," with others listed as "yellow" or "green."

These risk factors and the prioritized list of risks that result from them can be valuable tools. However, we need to be very careful and realize that the calculations are subjective. We need to be particularly careful with risks that have a low probability but a high impact.

For example, if our project involves a presentation by the governor, the risk of an assassination attempt is probably very low, although the impact on our event would be very high. That low probability and the resulting relatively low risk factor do not mean that we should not take necessary precautions. At a minimum, we should work with those who provide security for the governor to make the event as safe as possible. For risks with a very low probability but high impact, a slight change in the probability can dramatically change the risk factor. For that reason, we need to carefully monitor those risks, even though their risk factors do not warrant their placement high on the list of prioritized risks.

Table 11.3 Illustrative Risk Register

Risk #	Risk	Prob.	Imp.	Risk Factor (P X I)	Watch List	Risk Trigger	Risk Response Strategy	Secondary Risk

Two tools that allow us to continue to monitor risks are the identification of risk triggers and the assignment of risks to a watch list. Risk triggers are those events that signal that a risk is about to change or that a risk has occurred. For example, if one of the risks of the project were adverse press coverage, a risk trigger might be a request for an interview by a reporter known to be hostile to the current administration. When the trigger occurs, we need to reevaluate the probability of the risk and, potentially, implement a new strategy. We can identify some of the risk triggers in advance to help us monitor evolving risk conditions. A risk watch list simply allows us to identify some risks in advance that we want to note for special attention. Good candidates for risks to put on the watch list are those with a low probability but high impact and those that the team may have differences of opinion about. For example, if part of the team feels that the risk of a key vendor failing to meet a delivery date is a high probability, while other team members think it is a low probability, a compromise can be made by putting the risk on the watch list.

Other tools we can use for qualitative risk analysis are risk data quality assessment and risk urgency assessment. Data quality assessment analyzes the quality of the data used to score the risks for probability and impact. If the data is good, those scores will be more accurate. If the data is of low quality, the scoring cannot be accurate. Data quality assessment requires that we question our scoring and evaluate the data used for that scoring. Data quality assessment requires us to ask ourselves how we know the facts that drove the scoring.

Urgency assessment allows us to identify which risks require an immediate response and which risks can wait for analysis and response later in the project. Some risks require additional analysis. We can also use quantitative risk analysis, which is the statistical cousin of qualitative risk analysis. It is not applied in every project. The simplest example is a risk for which we need more information. For example, we may identify the risk that our project site may need to be evaluated and approved by an environmental protection agency. Investigation of the requirements of the environmental agency or a discussion with agency officials might quickly identify whether that risk applies to the project.

Quantitative risk analysis can also use advanced modeling tools like sensitivity analysis, decision trees, and computer modeling including Monte Carlo analysis. These methods, unlike the critical path method of schedule analysis, use probabilistic analyses (sometimes referred to as stochastic estimates) which do not identify a single activity duration, but rely on a range of potential durations or costs. Monte Carlo analysis operates by:

- Using a random selection process to select a duration for each activity that falls within the range of durations for that activity
- Calculating the length of the project for that set of selected durations
- Repeating that process many times
- Identifying the distribution of total project durations
- Computing the cumulative probabilities

With that distribution of results, the project team can calculate the probability of meeting selected dates for project completion. For example, the team can identify the date on which they can be 80 percent certain of completion. They can also identify how many days of buffer need to be added to the schedule to be 75 percent certain of completion. Of course, the accuracy of the Monte Carlo estimates is dependent on the accuracy of the duration estimates for each activity.

Developing Risk Responses

Risk response planning is the process of creating cost-effective responses to the identified and prioritized risks. Because it is unlikely that we will have the time or be able to develop cost-effective responses to all of the risks identified, we have to start with the most serious risks and develop responses that address the probability weighted impact. For example, if a risk has a 10 percent chance of occurring and could potentially cost the project $100,000, we would not want to spend more than $10,000 on a risk response ($.10 \times \$100,000$)

Although the focus of most projects is on negative risks, recall that the definition of a risk includes factors that could have either a positive or negative impact on the project. As a result, we need to identify strategies to reduce the impact and probability of negative outcomes *and* identify strategies that can increase the probability and impact of positive outcomes, which are sometimes referred to as opportunities.

We have two initial choices with regard to forming risk responses for our identified risks. First of all, we can choose to accept a risk. Though that strategy sounds overly passive, that is the option we will adopt for many of our risks. Many risks simply do not have cost-effective risk response strategies available. Particularly for those risks that are caused by factors outside our control (e.g. the potential actions of legislators or the courts), we may not have many options to reduce risks. If we accept risks, we can, however, keep an eye on them to identify how they might be impacting the project. For example, if there is a risk that federal rules will change in the middle of our project, we can monitor the federal rule-making process to determine the probability that the risk will occur.

For those risks we have accepted, we can also create contingency reserves. Contingency reserves are extra time or budget added to specific activities to account for risk that has no other response strategy. For example, if there was a chance that all state employees might be given a pay raise midway through the project, we could add a contingency reserve to the project budget. That contingency reserve is not a "slush" fund. It is there for a specific purpose, and the goal is to finish the project with the reserve intact unless the specific risk condition it was established for occurs.

For other risks, we can adopt strategies that can reduce either the probability or the impact of the risk, or both. In private-sector projects, purchasing insurance can reduce the impact of certain risks. Government organizations are often self-insured, which limits the applicability of insurance as a risk response strategy. Public-sector organizations can, however, create strategies with vendors that shift risk to them. For example, highway construction projects could include penalties for late completion, which transfers some schedule risk to the contractor.

Reducing the probability that a risk event occurs requires that we recognize the root causes of the risk and attempt to impact them. For example, if there is a risk that a key project team member will leave the project prior to its completion, we can analyze the factors that might cause the team member to leave and attempt to impact those factors. It might be that the team member does not feel that the work he or she is being assigned allows for personal growth. If that is the case, we could assign more interesting work to the team member to reduce the probability that he or she will look for other employment options.

Reducing the impact of a risk event on the project requires that we identify the potential impacts and buffer the project from the impact. For example, in the prior example we might conclude that the early departure of the team member would leave us without specific skill sets. To reduce the impact of the departure of the team member, we could build those skill sets in other team members or hire outside experts.

Another tool for managing the uncertainty of projects is the identification of project assumptions. Assumptions also deal with project uncertainty, but, unlike risks, project assumptions identify uncertainty and make a presumption about how that uncertainty will unfold. Assumptions state that, for the purposes of project planning, we are making a prediction about the future. With regard to risks, we identify a point of uncertainty but leave ourselves open to the possibility that the outcome is open to different potential outcomes. By clearly stating the assumptions that we are making in project planning, we are, in effect, identifying the circumstances under which the project can be successful.

For example, we might make an assumption that no legal impediments exist to the adoption of the new set of processes. By "vetting" that assumption with our stakeholders, we can get their agreement that the assumption is valid and appropriate for project planning. If it later turns out that the assumption was not valid, the project plan, including its schedule and scope, must change.

We need to be aware the any risk-response strategy creates secondary risks. Secondary risks are the risks that are created by the risk-response strategy. Secondary risks make it clear that, any time we respond to a risk by creating a risk-response strategy, we are trading off that risk for another. Our goal is to reduce the overall risk to the project and to trade off risks we are not comfortable with for those we are more tolerant of.

For example, if we believed that agency staff did not have the necessary skills required for a project, we could use a vendor instead of agency staff and reduce the risk that we do not have the necessary skills for accomplishing the project's goals. By adopting that strategy, however, we have created secondary risks that might include the risk that we will overspend the budget, the risk that the vendor may not deliver as promised, or the risk that we may become dependent on the vendor and never develop the skills in-house. Making smart trade-offs among risks is a challenge, which may not always have clear answers.

We can also develop contingent response strategies, which are designed for use only when certain events occur. For example, if there is a risk that the necessary personnel will not be hired on time, we could develop a contingent strategy to be put in place only when those personnel were not hired on time. In that case, our contingent strategy could be to contract with vendors to take over key project deliverables. Those vendor contracts would only be developed if the risk event occurred.

Last, we can identify residual risks. Those are the risks that remain after the deployment of our risk-response strategies. Developing risk responses in public-sector projects can be a particular challenge, because many of the most critical risks fall outside the control of the project. Public-sector project managers cannot control the press, the legislature, or even critical processes that they rely on.

Because of that, risk responses for public-sector projects are often less certain. Instead of taking steps to make certain that personnel are hired on time, the public-sector project manager may have no better strategy than to build good working relationships with process owners, make sure that senior managers are aware of needs, take action to comply with process requirements, make good estimates of the duration of resource use, and work with the organization to improve processes where the chance presents itself. These types of risk responses once again require the public-sector project manager to employ high-level communications and diplomatic skills.

When we have created our risk-response plans and assigned responsibility for implementing them, they are entered into the risk register. Like every other element of that risk register, our risk-response plans will require constant reevaluation and attention.

Unfortunately, risk management is not a "once-and-you're-done" set of processes. In a dynamic project environment, risks are fluid and demand constant attention. Some project managers even argue that project risk should be an agenda item at every project team meeting, and whether or not risk is discussed at every team meeting, risks must be continually monitored. Some risks will increase in probability or potential impact, and some risks will fall off the risk register. For example, we may have identified a risk as the potential that the purchasing agency will not select the vendor and finalize the contract to allow them to begin work by the required date. If that vendor is ready and able to begin work on the required date, the risk has disappeared. We can leave that risk on the risk register for historical purposes, but it is no longer of concern to the project team.

We will need to constantly update the risk register to reflect new risks, effective risk-response plans, adjusted risk-response plans, new assessments of risk probability and impact, assumptions that have been reevaluated and are now considered to be risks, new assessments of risk urgency, and changing risk triggers.

The activities we undertake to monitor risks and make changes to the project plan because of changing risk conditions are a part of the function described earlier as comparing the work to the plan and managing changes.

Creating a Plan for Managing the Project's Legal and Administrative Constraints

Every project in any sector is faced with legal and administrative constraints. Project teams have to be aware of the law, and no project is allowed to violate the law. Every project also faces administrative constraints. Even in private-sector projects, company policies and practices constrain the project. One of the best examples is the process employed to make payments to employees or vendors. Project managers need to be aware of those processes and conform to them if they hope to have project resources compensated on a timely basis. In the best case, however, these legal and administrative constraints simply represent the rules for project performance and do not unduly constrain the achievement of intended outcomes.

As we have repeatedly pointed out in this book, however, legal and administrative constraints are a major factor in public-sector projects and impact a much higher percentage of projects than the same kinds of constraints do in private-sector projects. They can:

- Stop the project dead in its tracks
- Add substantial amounts of time to the project schedule
- Cause project resources to be engaged in activities that do not directly produce project outcomes
- Widen the circle of project stakeholders to include legislators and oversight agencies
- Embroil the project team in costly, time-consuming legal battles
- Cause project team members to adopt overly conservative approaches to project work
- Require the utilization of high-priced consultants who are expert in legal and administrative issues

Responding to the constraints that can affect a public-sector project requires a four-step process similar to the functions required for general risk management. Those four steps are:

1. Identification of the legal and administrative constraints that might affect the project
2. Evaluation of their impact
3. Identification and evaluation of options for response
4. Selection of strategies for coping with these constraints

Identification of the legal and administrative constraints that might affect the project requires a broad-based process of inquiry that can include:

- Reviews of laws and administrative rules
- Examination of the experiences of prior projects
- Interviews with agency experts
- Interviews with oversight agency staff
- Consultation with legal staff

Once a list of potential legal and administrative constraints is compiled, the constraints can be evaluated for their potential impact. Some legal and administrative constraints can prevent a project from moving forward. Some states, for example, require all information systems to be centrally developed and maintained. An agency intending to build its own systems would be prohibited from doing so. These constraints that can prevent a project from being undertaken must be identified before the project commences. Discovering them after the project has been initiated can be very costly and very embarrassing for project sponsors or managers.

Some legal and administrative constraints have a lesser impact. Some merely require additional project activities to ensure compliance. Others may require the activities necessary for the amendment of rules or processes to be included in the project plan. In most cases, the development of a new public program will also require development of new administrative rules and all of the activities related to compliance with administrative process for rule development, which may include public hearings and participation in a rule-review process.

Next, the project team can identify and evaluate the options available for coping with legal and administrative constraints. In some cases, project plans can be changed to avoid the rule or process. It is harder to mitigate the impact of a law or rule since they are absolute in terms of compliance. That is, we cannot partially comply. We could, however, minimize the impact of purchasing rules by using in-house resources, to a larger degree, for the project. That, of course, creates additional project risks. Other options for dealing with legal and administrative constraints include:

- Engaging experts on working within the constraints
- Proposing alternative rules or laws and working for their passage

- Reducing the scope of the project
- Requesting waivers of rules if processes for waivers exist

Table 11.4 Template for Tracking and Managing Public-Sector Project Legal and Administrative Constraints

Constraint	Source	Potential Impact	Response Plan	Responsible Person

Having identified the options available for responding to the constraints, we can choose a set of strategies. Those strategies, of course, need to be integrated into the project plan. We can also use a format, such as the one illustrated in Table 11.4, for tracking and managing those constraints.

The source of the constraint should include the legal reference or administrative rule citation. Obviously, those constraints embedded in statute are the most critical, with those embedded in formally constituted administrative rules being second. Process constraints can sometimes be set aside and a waiver requested. Listing the source of the constraint allows for the identification of the seriousness of the constraint and the potential for changing or waiving the constraint.

In most project environments, laws or administrative rules are taken as a given, as the product of an uncontrollable set of political processes. Public-sector project managers, in some cases, are very close to the processes that create laws and rules. As a result, changing laws or rules may not be outside the realm of possibility for public-sector project managers. Making those changes will, however, require the involvement of a large set of stakeholders, including political players, as well as time and resources. The deliverables related to those changes should be included in the project's work breakdown structure.

BEST PRACTICES FOR PUBLIC-SECTOR PROJECT RISK MANAGEMENT

Best practices for managing risks for public-sector projects can include:

- Adopt a proactive strategy with regard to risk
- Engage as broad a group as possible in risk identification and response planning

- Identify the risk tolerances of your organization
- Keep in mind that risk can be reduced but not eliminated; be alert to secondary risks
- Give special attention to those points in the project at which control passes out of the hands of the project team and into the hands of other offices or processes
- Create risk categories that include political and media risks and the risk of failing to deliver a product that will satisfy customers
- Use risk analysis methods that fit the project and the organization (i.e., methods that are neither too rigorous nor too informal)
- Put the consideration of risk on the agenda at every team meeting
- Assign responsibility for the management of specific risks
- Build a plan for managing legal and administrative constraints
- Include time and resources in the project plan for the implementation of the plan for managing legal and administrative constraints

DISCUSSION QUESTIONS

1. What other categories of project risk can you think of for public-sector projects?
2. What challenges have you found in managing risks for public-sector projects? What risk management processes have you employed? What risks specific to public-sector projects have you experienced?
3. What methods do you use for risk monitoring and control? Do you actively engage in risk management throughout the project, or is it a one-time analysis?
4. What have you discovered about stakeholder risk tolerances and preferences in public-sector projects?
5. What special challenges have you faced in managing the legal and administrative constraints of your public-sector projects?
6. What other best practices can you imagine for managing the risks of public-sector projects?

EXERCISES

1. For a project you are familiar with, create a risk-impact scale for negative risks that fits the risk factors and tolerances of your organization and another for the positive risks.
2. For a public-sector project you are familiar with, identify a set of project risks. Use the risk categories you identified in the risk management plan to help with your identification.
3. For the list of risks that you identified in the last exercise, assign a probability and impact score and multiply the two to create a risk factor. In addition, identify risk triggers for the most significant risks, and designate some of the risks for inclusion in the watch list. Use the risk register template provided.
4. For the five most serious negative risks that you identified earlier, develop risk-response strategies. Include more detail than saying, "this risk will be mitigated." Instead, identify how it would be mitigated, transferred, avoided, or accepted. Identify who would be responsible for managing the risk-response strategy.
5. Develop a plan using the template illustrated in Table 11.4 for managing the legal and administrative constraints of a public-sector project.

Closing Willowbrook

The Willowbrook State School on Staten Island in New York State hardly lived up to its pastoral name. Instead, it was a warehouse for more than 6,000 people with mental retardation and was well over its stated capacity for 4,000 residents. Senator Robert Kennedy visited Willowbrook and called it a "snakepit" even before its budget was slashed in the early 1970s. Although some parents and advocates had tried to draw attention to the conditions at Willowbrook, nothing changed until a young TV reporter named Geraldo Rivera smuggled a handheld camera into one of the wards and ran his tape on the evening news in January 1972.

(continued)

The energy created by Rivera's report resulted in a suit being filed by the American Civil Liberties Union (ACLU) on behalf of the residents. The new Governor of New York, Hugh Carey, quickly signed a consent decree rather than engage in lengthy litigation that the State would likely lose. That consent decree mandated that, instead of pouring millions into Willowbrook and creating a "shinier prison," Willowbrook would cease to exist. Thus began the largest single effort at deinstitutionalizing people with mental retardation in the nation.

Initially, Federal Judge Orin Judd oversaw the implementation of the first consent decree and its successor decree. When Judge Judd passed away, oversight passed to Judge John Bartels. Day-to-day oversight of the actions of the State Office of Mental Retardation and Developmental Disabilities (OMRDD) was managed by the Office of the Special Master, an appointee of the federal court, who had the responsibility of tracking the progress of the State and reporting to the court.

The challenge for OMRDD was to find or build community facilities for the Willowbrook residents and those who were on waiting lists to be sent to Willowbrook at the time of the consent decree. Over the years, Judge Bartels issued more than 50 orders and opinions and held more than 100 hearings to prod OMRDD to comply with the order signed by the governor.

In 1987, a closing ceremony was held at Willowbrook, but many former residents were still in other State institutions. By 1992, all but about 150 members of the Willowbrook "class," those covered by the consent decree, had been moved to community facilities.

In 1993, Judge Bartels, then 95, formally released the State from oversight and ended the Willowbrook saga, though the State entered into another agreement that guaranteed that Willowbrook residents would forever remain in community facilities.

Chapter 12

Managing Project Procurement and Vendors

This chapter examines the methods for managing procurement in public-sector projects and takes a broader look at how outcomes can be optimized when vendors are involved. That broader outlook is necessary, because managing vendors takes both a solid set of processes and a different management style.

Public-sector purchasing processes are more complex than their private-sector counterparts and have additional objectives that include:

- Maintaining fair processes
- Avoiding conflicts of interest
- Establishing an objective basis for vendor selection
- Obtaining the best value
- Documenting seller performance and requirements
- Obtaining flexible arrangements

Today, public-sector organizations are experimenting with a variety of creative relationships with vendors and, even in some cases, employees as they experiment with programs that allow employees to work from home or remote sites and build their schedules using flexible hours. In this new organizational environment, the role of managers has changed. Rather than only managing long-term, on-site employees, whose activities are monitored on a daily basis, managers in today's modern public-sector organization and those managing public-sector projects are being asked to:

- Manage resources with only periodic, rather than daily, personal contact

- Devise creative yet accountable schedules and work arrangements, such as job sharing and less-than-full-time work
- Manage vendors who have been assigned critical organizational functions
- Make the most of periodic face-to-face meetings and even-rarer full-team meetings
- Communicate more effectively using electronic technologies

Meeting these challenges requires a new set of management skills, which are focused more on managing toward outcomes rather than activities. The challenges include identifying those outcomes; specifying them clearly; managing changes in needed outcomes; creating the legal frameworks that build relationships; ensuring compliance with laws, rules, and policies; and managing transitions.

This chapter looks at how public-sector project managers can meet both the challenges of public-sector project procurement and the challenges of the new workplace environment. In specific, it details the required functions of project procurement management; describes the challenges of vendor, outcome, and performance management; identifies the legal framework for outsourcing and vendor management; explores the development of effective service-level agreements and statements of work; and identifies methods and tools for managing changes in vendor relationships

The intention of this chapter is not to recommend outsourcing of services or products in public-sector projects. Instead, it is intended to recognize the inevitability of the new environment and prepare public-sector project managers to deliver successful projects even when those projects require the use of vendors.

THE NECESSARY FUNCTIONS OF PUBLIC-SECTOR PROJECT PROCUREMENT MANAGEMENT

Earlier in this book, three necessary functions for project procurement management in the public sector were identified. They were:

- Identifying necessary purchases and resource acquisition
- Working with purchasing offices to identify and select vendors
- Managing contracts and vendors

Each of these functions is described in turn in the following sections.

Identifying Necessary Purchases and Resource Acquisition

Even though the "how" of purchasing in public-sector projects is clearly defined by purchasing processes that may have their requirements embedded in statutes, public-sector project managers still have to consider what they will purchase for their projects. Although purchasing equipment and supplies can be a challenge, the most challenging decisions are those that identify how human resources will be acquired.

Make-or-buy decisions require the analysis of whether the goods or services can be better created in-house or through the use of vendors. Some decisions are easy. It probably is not efficient to build computers in-house from purchased parts. On the other hand, creating in-house software versus buying off-the-shelf or custom-built software is a far trickier question. There is also no simple answer to the question of whether to use vendors or in-house resources for projects.

In the private sector, quantitative, financial criteria can be applied to those decisions, employing structured make-or-buy analysis. Even in the private sector, however, other criteria can also apply to the decision to use in-house resources rather than to purchase resources from outside the project or contract for services with vendors. Some of those other criteria can include consideration of:

- The willingness of the organization to share project data with the vendor
- The organization's long-term need for capacity development (Vendors may be cheaper than using in-house staff, but vendors typically take their knowledge with them when they go.)
- The need for secrecy
- A philosophy of allowing in-house staff to grow in their jobs
- A prior history of using vendors successfully (or the reverse)
- The availability of vendors with the necessary skills
- The administrative costs and time needed for hiring vendors versus the administrative costs and time needed for hiring staff
- The need for the good or service beyond the life of the project
- The availability or nonavailability of in-house staff
- Internal Revenue Service (IRS) rules

In addition to these nonfinancial criteria applied by private-sector organizations as they consider the use of in-house resources versus vendors, public-sector organizations also have to consider other factors. Those additional criteria include:

- Legal requirements and restrictions on the use of vendors
- Political pressures
- Union or civil service requirements
- Media oversight
- Purchasing processes that attempt to address social goals

Whether or not public-sector organizations employ these nonfinancial criteria, at least some portion of the purchasing decision should rely on make-or-buy analysis. That analysis requires the comparison of the costs of using in-house resources with the costs of acquiring them from outside. The costs of in-house resources can include:

- Salaries and benefits
- Indirect costs charged to the project
- Training to ensure that employees have the necessary skills
- Management and supervisory costs (When some jobs are outsourced, the management of the resources providing services may be the responsibility of the vendor.)

The costs of acquiring resources for the project from outside the organization can include:

- Direct costs paid to the vendor
- The costs of the acquisition process
- The costs of contract management
- The potential costs of legal action
- Long-term vendor costs if in-house capabilities are not developed

Another factor that may influence make-or-buy decisions is any productivity differential that may be perceived to exist between in-house staff and vendors. In some cases, in-house staff will simply not have the necessary skills for the project. In other cases, in-house staff may have the necessary skills but may not be able to perform those skills as efficiently as vendor staff might. That is sometimes the case when vendors have more experience in the application of sophisticated or proprietary information technologies. Those productivity differentials need to be weighed against cost and nonfinancial criteria, such as those listed earlier, and can be factored into make-or-buy analysis.

If the project team elects to acquire human resources from outside the agency, it will also need to determine if it intends to hire those external resources on a part-time or full-time basis, or if it intends to issue a contract and treat them as independent contractors. The Internal Revenue Service (IRS) identifies rules for treating a resource as a staff person or a contractor. Those rules are biased toward employment rather than contracting to ensure that withholding taxes are paid by the employer rather than passed on as an obligation to the contractor. Public-sector project managers should always consult with human resource or legal experts when they are faced with this decision, because substantial penalties can accrue to the organization if the IRS determines that contractors should have been classified as employees.

Once the project team has considered the resources it needs and has completed the necessary make-or-buy analyses, it can create a list of resources to be purchased or acquired from outside the organization. With that list in hand, the project team can begin to create a plan for managing the purchasing and acquisition processes.

Working with Purchasing Offices to Identify and Select Vendors

In public-sector organizations, the responsibility for purchasing goods and services are usually carefully defined and assigned to designated purchasing officers or departments. In addition, most public-sector purchasing processes are well-defined and represent a constraint for the project. Those constraints can affect the project schedule, budget, risk, and quality.

Public-sector purchasing processes are intended to:

- Ensure that the best price is achieved, including making use of bulk-purchase discounts
- Limit opportunities for conflict of interest or outright theft by agency staff, avoid politicization of the purchasing process, and limit conflicts of interest and the potential for ethical conflicts
- Accomplish social goals, like providing for more minority contractors

Although the project team cannot directly control purchasing and resource acquisition processes, public-sector project teams need to create a plan for managing those processes and interfacing with procurement agencies so that necessary goods and services can be purchased and the goals of the project met. That plan could contain such elements as:

- Identification of purchasing processes and rules that may constrain the project
- Team members responsible for working with purchasing agencies
- Identification of procurement documents that the project team will need to complete
- Methods, if available, for expediting purchases and resource acquisition
- Methods to be used for vendor selection
- Points at which project managers can influence purchasing decisions
- Durations of purchasing processes and their impact on the project schedule
- The types of contracts that can be used

For public-sector projects, the procurement plan might also include the ethics and conflict-of-interest provisions that apply to the project team and to vendors. For example, public-sector employees are usually prohibited from accepting anything "of value" from vendors or outside parties. There are different definitions of what something "of value" constitutes. In some cases, it can be as small as a cup of coffee. In other jurisdictions, a dollar limit is placed on daily gifts.

All project team members must be aware of those rules, and the project manager should monitor compliance. Few things can derail a public-sector project faster and more thoroughly than ethics violations or even the allegation or appearance of ethical violations. In addition, public-sector vendor management plans should ensure that both vendors and agency staff clearly understand the limits of their authority and the mechanisms for creating changes and accepting deliverables.

Public-sector project managers also have to be aware of the influence that economic or market conditions have on public-sector purchasing systems. If markets for required goods and services are competitive, the procurement process will likely require that a competitive process be applied. In those circumstances in which vendor competition does not exist, sole-source purchasing can often be applied. Vendors will sometimes attempt to convince purchasing officials that they offer a unique product or service to prevent the imposition of competitive bidding processes. Competitive bidding processes typically add time to the project but can result in lower costs if a productivity differential between the best and the cheapest provider of services does not eat up savings. The project team needs to create a workable balance between hiring vendors that they are comfortable with against the cost benefits of competitively bidding project work.

In some cases, waivers of competitive bidding can be applied if the dollar amount of the contract falls below a preset level.[1]

Once the project team has decided what to buy and created a plan for managing the interfaces with purchasing agencies, another question to address is the type of contract to employ. Three general types of contracts exist:

- *Fixed-price contracts*, which pay an agreed-upon price for goods or services without regard to the cost incurred by the vendor
- *Cost reimbursement contracts*, which reimburse the vendor for legitimately incurred costs of providing services or good
- *Time and materials contracts*, which pay an agreed-upon hourly rate for services, which is multiplied by the number of hours provided

Other variants can be created to provide incentives for vendor performance. Payment incentives can be provided to vendors for coming in under cost targets or for complying with other standards. For example, many public-sector road-building contracts contain incentives for early completion.

It is commonly believed that fixed-price contracts are to be employed if the buyers of goods or services are uncomfortable with price risk—the risk that the price of the good or service could vary dependent on the costs incurred by the vendor. That can be a concern.

Of more importance, however, is the extent to which the buyer of goods or services can define the product to be produced. If the buyer (the public agency) can clearly define the scope of the goods or services to be purchased, it can use a fixed-price contract to shift the price risk to the vendor.

---

[1] Project Management Institute, *Government Extension to the PMBOK® Guide Third Edition*, 2006, p. 73.

That way, the buyer does not have to worry or care about the costs incurred. That risk belongs to the vendor.

If the scope of goods or services is not clear, or if it changes in the course of the project, several problems can arise if a fixed-price contract is in place. For example, the vendor can allege that the product they have provided a fixed price to deliver is not the product being asked for later. In that case, the vendor can argue that the fixed price is no longer valid and that new costs have been incurred. Similarly, it is possible that vendors working under fixed-price contracts would argue that components or work to be completed by agency staff are unacceptable and that new costs need to be incurred.

An agency is under a substantial disadvantage if new deliverables have to be added to a fixed-price contract, irrespective of the cause of the dispute or reason for new costs, in that the fixed-price contract is no longer valid. The vendor holds the upper hand in negotiations, unless the agency is willing to terminate existing contracts and seek a new vendor. In most cases, terminating a vendor would be costly, because it would incur the costs of finding a new vendor and, in many cases, the agency would need to pay penalties to the former vendor. At a minimum, costs incurred to date would likely need to be paid, even though deliverables may not have been completed.

In some fixed-price engagements, an agency may be engaged by the vendor in discussions of what the definition of goods or services is. The vendor will have a built-in incentive to minimize the definition of the product or services to reduce its cost and increase the margin on the deal. The agency has an incentive to maximize the definition of goods or services to get the most from the vendor for the price paid.

Cost reimbursement contracts have their own problems. Under cost reimbursement contracts, debates can arise over the definition of legitimately incurred costs. Agencies can stipulate in advance what costs are eligible for reimbursement, which may include both direct and indirect costs.

The U.S. Federal government, for example, establishes preapproved annual indirect cost rates for some providers of services (e.g., universities). Those preapproved rates are created through a process that requires detailed cost data from the vendor and an audit of that data by a cognizant audit agency of the government.

Debates will still often arise about legitimate costs. For example, what happens if the vendor assigns staff to a project who are not competent and do not add value to the attainment of project goals? Clearly, the costs of personnel assigned by the vendor to the project are reimbursable costs of performing services. But what happens if assigned staff persons are not capable and do not contribute to project outcomes. (Some public-sector purchasing contracts attempt to mitigate this risk by allowing the agency to preapprove vendor staff before they are assigned to a contract.) Public-sector project managers should always seek the advice and assistance of agency legal staff as they attempt to enforce contract terms.

Public-sector project managers need to create a good relationship with procurement agencies so that they can serve as an agent of the project, though they will always be constrained by their responsibility to ensure that laws and rules are complied with. In order to ensure that agency procurement processes create results that optimize the possibility that the project will be successful, the project team needs to work closely with those processes to:

- Develop the list of qualified sellers to which the bid documents will be sent
- Prepare the final bid package and transmit it to the potential vendors
- Receive the seller-prepared proposals
- Evaluate vendor proposals
- Select the best vendor
- Develop the contract

In public-sector projects, the list of qualified sellers may have been prepared in advance of the project by the purchasing department. In addition, that department may assume full responsibility for preparing the bid package and receiving the proposals. One challenge for public-sector project teams is influencing that process, in appropriate ways, to ensure that the vendors selected can deliver goods or services that contribute to the project. In the best circumstances, the purchasing department is an ally of the project team.

Those purchasing agencies will also usually issue the procurement documents, including requests for proposals, contracts, and statements of work. Procurement documents are sometimes referred to as requests for proposals (RFPs), invitations to bid, or requests for quotations (RFQs). Project managers must ensure that they have input into the preparation of those documents, so that the best vendors can be selected. Purchasing offices sometimes have a bias toward choosing the lowest-cost vendor, a bias that the project team must try to prevent. That bias may have been created as purchasing agencies try to counter the desires of project managers to hire the best resource irrespective the cost. In an earlier chapter, we described the failure of some public-sector projects to capture and manage costs. If costs were tracked and managed and the project manager was held accountable for cost management, the need for the purchasing agency bias might be reduced.

Procurement documents can contain several elements, including:

- The statement of work to be performed
- Minimum requirements for vendors
- Due dates for proposals
- The required format of the response
- Pricing requirements, which can include such considerations as the requirement to provide hourly rates for assigned vendor staff, a total fixed cost, overhead rates to be applied, costs per project phase, and other cost detail

The project team also needs to work with the purchasing agency to identify the evaluation criteria to be applied to vendor proposals. The evaluation criteria to be applied to public-sector vendor bids usually are composed of three components:

1. *Compliance with preregistration processes*, which allow only those vendors who are registered with the jurisdiction to submit bids
2. *Screening criteria*, which are used to determine whether vendors meet the minimum requirements
3. *Evaluation criteria*, which allow project team members and purchasing organization staff to evaluate the relative costs and benefits of each vendor's response

Precertification processes for vendors are applied by many public-sector organizations. Those processes required vendors to meet minimum standards to be included on a bidder's list for agencies. Those processes also often require vendors to offer their best and lowest price to agencies. Precertification of vendors can also expedite the purchasing process by limiting the list of vendors eligible to bid and having established in advance that they meet minimum criteria.

Procurement documents typically include minimum criteria for selection. Those minimum criteria, sometimes called screening criteria, allow a "first-cut" evaluation of the proposals to eliminate vendors who do not meet standards. Minimum requirements for vendors for public-sector items can include such things as:

- Licenses to do business in the jurisdiction
- Requirements for having been in business for a minimum period of time
- Conformance with social goals. such as having affirmative action plans
- Identification of minority business status, which may not disqualify vendors but which might give them preference if they meet the definition
- Business insurance or a performance bond
- Vendor number assigned to those who have met prequalification requirements

The vendor's response to procurement documents, including their responses to minimum requirements, can be incorporated into the contract as an appendix. That requires vendors to make a cost estimate that they are willing to adhere to, and also attaches their answers to minimum requirements to the contract. If the vendor is later determined to have misstated elements of the minimum requirements, the contract can be easily voided.

For example, a political jurisdiction in Ohio required vendors to sign a statement as part of the minimum requirements for a bid on public-sector contract work that indicated that the vendor had not sold weapons to militants in Northern Ireland. One might argue that such a statement was useless in that, if a vendor had sold weapons to militants in Northern Ireland, which is against the law, they would have lied about it in their bid response. Although that might be true, if the vendor were later determined to have sold weapons to militants, the contract could have been voided on the basis of false representation.

Evaluation criteria should be specific to each need for a vendor, and public-sector agencies have employed a wide array of evaluation criteria. They can include such elements as:

- Vendor costs, though, in some cases, costs are considered separately and are required to be submitted in a separate document from the other components of the bid

- The vendor's approach to the work as described in the bid
- The qualifications of the vendor's assigned staff
- Similar work done by the vendor in the past, including work for similar agencies
- The vendor's familiarity with the agency and its environment
- The vendor's status as a minority vendor or women-owned business, which can be used to provide additional points in an evaluation system

To be applied most effectively, weights should be assigned to each evaluation criteria, and several evaluators should independently review the bids. When all of the evaluators have completed their review and scoring, the scores can be compared to determine the overall winner. Using independent evaluations in this manner can eliminate some of the subjectivity that can accompany the evaluation of vendor submissions.

Although subjectivity can be reduced by using detailed evaluation criteria, project team members should keep in mind that the choice of evaluation criteria can also bias the selection toward one vendor. For example, if a high weight were assigned to "prior work with the agency," those vendors who had worked with the agency in the past would be given preference over new vendors. Although public-sector project teams may want to hire vendors they have worked with in the past in order to reduce the vendor learning curve, they must also keep in mind that new vendors may have different approaches and the ability to add new value to the project.

Documentation of the selection process and the scoring should be maintained in the event that a vendor challenges the results and alleges bias in the selection process. Public-sector purchasing processes may also require that bids be advertised, using so-called mastheads or other notices in newspapers. Advertising can help expand the list of vendors willing to do the work.

In some cases, a bidders' conference is required as part of the issuance of the bid documents. Whether one is required or not, a bidders' conference can be a valuable tool or the project team. A bidders' conference can:

- Require attendance by vendors as a means of gauging the number of interested vendors
- Ensure that all bidders receive the same information
- Cut down on the attempts by potential vendors to elicit information about the project and the selection process

The bidders' conference can also help the project manager and agency enforce ethics rules. By forcing all communications with the vendor to take place through the bidders' conference, the agency ensures open and fair communications and prevents the development of an inside track to the engagement. If the bidders' conference requires mandatory attendance by potential vendors, it also allows vendors to assess the competition for the engagement. Whether or not there is a bidders' conference, a public-sector project team must ensure that all potential vendors receive the same information and that answers to questions are shared with all vendors. Some public-sector agencies require potential vendors to submit their questions in writing and provide written answers to all of the potential vendors. Some public-sector agencies also require potential vendors to provide a letter of intention to bid before the actual bid submission as a means of assessing the number of vendors who are likely to respond to the RFPs.

Once bids have been received, in order to select the best vendor, project teams can apply a stepwise evaluation consisting of:

- Using the screening criteria to eliminate vendors who do not meet the minimum criteria
- Using weighting systems to score the remaining vendors
- Gathering independent estimates of cost and duration as a means of checking on the estimates provided by the vendors
- Soliciting information from those in the agency who have worked with the vendors in the past
- Requiring presentations from several finalists

When that process is complete, the project team can announce the selected vendor and engage in contract negotiations. In some cases, more than one vendor will be selected in order to ensure that the agency does not become reliant on a single vendor. Procurement documents often note that more than one vendor may be selected and that portions of the proposal from a vendor may be selected and combined with the work of other vendors.

As soon as possible, the vendors who have not been selected should also be informed that they were not chosen. In some cases, agencies may delay notifying all of the nonselected vendors in case the contract negotiations with the selected vendor break down. In most cases, the agency's purchasing department will have processes in place for vendor notification.

Contract negotiations for public-sector projects are usually not handled by the project team. Instead, agency legal counsel is responsible for contract negotiations, and the authority to enter into a contract is usually narrowly defined by statute. The project team will need to work closely with legal counsel to ensure that contract provisions do not unduly impede project work and that deliverables specified in the contract are appropriate for the project.

Managing the Contract and Vendors

Once a vendor is chosen and the contract is in place, management of the vendor's performance typically falls to the project manager and project team. Activities they may be required to engage in to manage vendor performance include:

- Monitoring vendor performance and compliance with agency work rules and practices
- Ensuring that necessary information is shared with the vendor
- Accepting and inspecting deliverables
- Ensuring integration of the vendor and team
- Evaluating proposed changes
- Approving payments
- Managing conflict within the parameters defined by the contract
- Working with legal staff
- Determining the need to escalate issues
- Providing feedback to vendor managers on assigned resource performance
- Ensuring the maintenance of contract records

For public-sector projects, project team members need to ensure that they understand the terms of the contract and their roles. Although project team members may prefer to operate informally with vendors, they need to be very careful to ensure that no actions of the project team have compromised the terms and operation of the contract. Project team members should be especially careful in the early stages of a vendor relationship when the risk of problems is highest.

Part of contract management is contract closeout. Although some contracts seem to be extended repeatedly, all contracts eventually reach their end. Ultimately, if all of the deliverables have been handed off, and if final payments have been made, the contract must be closed. Contracts can also be terminated early. Most contracts will define the circumstances under which they can be terminated. Contracts can specify termination:

- At the convenience of one or more of the parties with adequate notice
- Upon breach of the contract terms by one or more of the parties

For creating and managing contracts related to public-sector projects, legal counsel should always be engaged. That is also the case in contract termination. Even if the contract has been completed, legal counsel should still be notified so that appropriate notice can be given to the vendor.

Contract documentation must be preserved in case either party challenges a contract outcome. Each jurisdiction will specify what documentation must be preserved and for how long. The project team should also identify lessons learned and submit information about its experience with the vendor to the in-house database of vendor evaluations, if one exists.

The three required functions detailed for managing public-sector project procurement provide a solid basis for planning project procurement, selecting vendors, and managing contracts. But in order to maximize the contributions by external resources to the project, a new mindset and new skills are required. Those requirements for success are considered in the next several sections of this chapter.

THE NEW DEMANDS ON MANAGERS AND NEW TOOLS FOR MANAGERS

As noted early in this book, a fast-paced economy and worldwide competition require new organizational solutions, and, though the public sector lags behind the private sector in the utilization of these solutions, the public sector has begun to explore them and will, likely, increasingly rely on them in the near future. These new solutions rely more on outsourcing of goods and services than in the past.

In this emerging environment, public-sector managers will be challenged in ways that they have not been challenged before. Some of the challenges they face include:

- Motivating employees who are coping with increasing demands
- Dealing with a multigenerational workplace (According to some observers, there is a wider age range among employees in the workplace now than at any time in history.)

- Managing for short-term results with limited resources
- Managing employees who are not in the same geographic location
- Managing vendors who are now performing critical organizational functions
- Because of the increasing potential for layoffs in the public-sector, building organizational loyalty without the trade-off of a guarantee of long-term employment
- Supervising technical staff who have knowledge and skills the manager does not have
- Managing in an environment of constant change

So what are managers to do? One tool that will help public-sector managers, and that is specifically applicable to the management of vendors, is the management of outcomes. Indeed, it can be argued that project management is a deliberate attempt to hold employees and vendors accountable for creating outcomes.

Outcome management is a replacement for the management of activities. Although managers are being forced to adopt outcome management because of the workplace changes and reliance on vendors described earlier, it is actually a better strategy for managing the workplace and achieving results than micromanagement of employee activities. Outcome management is the only viable management tool available with vendors and off-site employees, since their activities cannot be easily monitored.

THE DIFFERENCES AMONG ACTIVITIES, OUTPUTS, AND OUTCOMES

In order to understand the different management strategies we need to apply to the management of outsourced providers or distant employees in public-sector organizations and projects, we need to understand the key terminology of performance management: inputs, activities, outputs, and outcomes.

Inputs are the resources used to produce outputs and outcomes. Inputs to projects include person hours of labor, resources like electricity and gasoline, or any other resource we need to do our work. Inputs to vendors performing work for our projects include cash payments, information provided by the project team, and work products provided to them.

Activities are the things we do in order to perform our jobs. In the language of project management, they are the verbs of our efforts. Activities can include such things as:

- Entering data
- Filling out a form
- Writing lines of code
- Completing a report
- Attending a meeting
- Making a presentation

Activities are what keep employees busy. Activities are easy to manage because they are observable and measurable. Unfortunately, many managers and management systems are focused too heavily on activity measurement, which tell us how busy employees are but not how productive they are. We place too much value on indicators of activity like:

- What time employees arrive at work and leave
- How busy they appear to be
- How many things they can document having done
- How many meetings they attended
- How full their calendars are

With full-time, on-site employees, managers can attempt to make a clear connection between required activities and the desired outputs and outcomes. Activities can be observed and tracked on a daily basis to ensure that outcomes are produced. Some managers, particularly those who might be guilty of micromanagement, prefer activity management. Activity management is also a tendency for managers who have been promoted from prior roles as subject management experts. Those managers have a tendency to rely on management that requires employees to "do it my way."

When we manage vendors or distant partners, we cannot manage their activities. Only in rare instances will we define the activities that vendors need to perform in order to meet organizational interests. Activities are typically left to the vendor to define. That is particularly true for fixed-price contracts but also may be true in cost-reimbursement contracts. If we cannot manage activities in outsourced arrangements, we have to make a choice of managing outputs or outcomes.

Outputs are what the organization produces. They are the completed products created by organizational activities. Outputs can also be easily measured and observed. Examples of outputs for public-sector organizations include:

- Number of permits issued
- Number of applications reviewed
- Number of accounting reports created
- Number of projects completed
- Number of customers served

Outputs are more critical to the organization than activities, but they still do not get to the ultimate outcomes of the organization. For example, the ultimate goal of a public-sector agency responsible for natural resources is not the issuance of hunting and fishing licenses. The goal of the agency is protecting the environment, and issuing licenses is an outcome that may contribute to that goal but is not the goal itself.

Outcomes are things that are of direct consequence to internal or external customers. They can also include indications of service quality as measured by customers. Although we do not think of it this way, we are employing outcome management when we buy any product from a vendor. For example, when we buy a loaf of bread, we rarely inquire as to what activities the baker engaged in to create the loaf. We do not care if he baked the bread in a large lot or a small one. Nor do we care how he managed his employees. All we care about is the outcome—the price of the loaf of bread and its ability to satisfy our needs.

End outcomes for public-service agencies are the end result that the organization seeks to achieve. End outcomes can include the accomplishment of social goals or the efficient allocation of public resources. In the public sector, outcomes usually are measured by how they affect the public interest and establish a fair and healthy society. In the prior example of an agency responsible for natural resources, outcomes could include:

- Preservation of natural resources for future generations
- Public enjoyment of natural resources in a manner that does not harm those natural resources
- Attraction of tourists to the jurisdiction and the generation of business and tax revenue

- Compliance with the requirements of statute
- Responsible use of natural resources in business and industry

The achievement of those outcomes is far more important than the activities the agency undertakes or the outputs of the agency.

THE CHALLENGES OF OUTCOME MANAGEMENT FOR CONTRACTORS AND VENDORS

It is clear that the measurement of activities will not produce good results for portions of a public-sector project outsourced to a vendor. The disadvantages of activity management include:

- The risk that the wrong activities will be performed
- The risk that the outsourced service provider can appear to be busy without creating effective outcomes
- The requirement that the contracting organization know exactly how work should be performed in order to specify activities
- The requirement that the contracting organization oversee the work (rather than the work products) of the vendor

In outsourced arrangements, activities cannot be observed, and, in many cases, the service provider has far more knowledge of the tasks to be performed than the organization contracting for the services. If they did not have that knowledge, we would not have hired them.

Because of the risks inherent in activity management, outsourced arrangements and vendor management require output and outcome management. There are challenges, however, with output and outcome management. Those challenges include:

- Identifying outcomes that may not have been well-defined by the organization
- Converting the organization from an activity management to an outcome management culture
- Ensuring that the right outcomes and outputs have been identified
- Developing methods for measuring outputs and outcomes
- Specifying outcomes and outputs in adequate detail to allow compliance with contracts and vendor performance to be measured

Project management forces us to make that shift to outcome management. One method for applying outcome management is the assignment of deliverables to in-house staff or vendors. If we believe that those to whom we assign those deliverables have the capability to create them, we can allow them to define their own activities. Another tool for moving toward outcome management is performance management, which is described in the next section.

PERFORMANCE MANAGEMENT

The process of developing an outsourced service relationship or a plan for managing an off-site employee requires the development of a performance management system. That system requires that we:

- Identify the current state of affairs
- Identify the desired state
- Identify the difference
- Create a plan to close that gap

That gap is sometimes referred to as a performance improvement zone. As we create a work plan for a vendor in a public-sector project, we might identify a current state in which we have the internal capability during the term of the project to create seven specific software modules of the total of eleven required for the project. The desired state is the delivery of all eleven modules at levels of quality that satisfy the project and on dates that meet the project schedule. The gap is the creation of four modules. Our contract with the vendor and the resulting service-level agreement becomes the plan for delivering the four modules and closing the gap between what we have the capability to create and our need.

Performance management can also be applied to project team members. We could identify the current state of affairs with regard to the employee's performance, compare that performance to the desired state, and, with the assistance of human resource staff, create a performance improvement plan for the employee. That plan, in order to be effective, would require performance metrics so that both the employee and managers could know without a doubt that the goals had been met.

We can employ a variety of tools to identify the existing state of affairs. Those tools include:

- Process mapping
- Customer surveys
- Benchmarking
- Identification of performance metrics
- Requirements analysis of the types described in the chapter on project quality management
- Make-versus-buy analysis

Table 12.1 Performance Planning Template

Current state of affairs	**Desired state of affairs**	**Gap**	**Performance improvement plan (measurable outcomes)**

A template for creating a performance management plan is included in Table 12.1.

MANAGING THE CULTURAL CHANGES NECESSARY FOR SUCCESSFULLY MANAGING VENDORS

Making the move to outsourced services, project vendors, or even distant employees and partners requires cultural as well as management changes. Those cultural changes can be as difficult as the operational challenges.

Some of the cultural changes necessary are:

- Helping managers understand that providers operating under contract may not be as responsive to crises as employees
- Helping managers understand that employees may not have the means to meet requests, particularly if processes and data are under the vendor's control

- Helping managers understand that requests for information or performance will have to be preplanned or anticipated
- Making the change from managing activities to managing outputs and outcomes
- Building trust within the constraints of contracts rather than building trust over years with colleagues
- Working through cultural differences that may extend across ethnic groups and nations
- Working with employees who may not want to give up functions or data

Managing the cultural changes and the emotions that accompany outsourcing may be harder than creating contracts and service-level agreements. To create those cultural changes, there is no substitute for vigorous and frequent dialogue.

THE LEGAL FRAMEWORK FOR OUTSOURCING PROJECT PRODUCTS AND SERVICES TO VENDORS

Even when internal employees were used for services, the relationship between the employer and the employee was governed by a legal framework. In this section, the legal frameworks applied to outsourced services are examined.

Why is it difficult to create prenuptial agreements? Although this sounds like a joke, the problem with creating prenuptial agreements is that good intentions to build a lasting, close relationship are mixed with legal issues. The same thing is true of outsourcing.

Although the goal is to create a win-win relationship with the vendor that can grow over time and blossom into a true partnership, that relationship has to be defined with legal documentation.

Most outsourcing arrangements consist of two fundamental documents:

- *The contract*, which is the basic governing document of the agreement. Any other documents are subordinate to the contract. The advice and assistance of an attorney is required for creating the contract.
- *The service-level agreement*, which defines in greater detail the services to be performed. The service-level agreement is examined in the next section. The service-level agreement is sometimes also

called the statement of work, though a statement of work is more often used to describe physical items to be purchased. The service-level agreement is designed to allow easier changes than the contract.

Both documents are designed to form the basis of the relationship between the parties. Therefore, they have to balance flexibility with accountability and clarity. That is a challenge that some organizations fail to meet. Keep in mind that vendors, when asked to perform duties not clearly specified in the documents, will have to fit those extra duties into the agreement and determine what constraints the agreements create. That requires a culture shift in that, before outsourcing, management could have asked full-time employees to just work harder or work overtime.

Elements of the Outsourcing Contract

Most outsourcing contracts will contain some combination of the following elements: (The definitions cited here are taken from general sources and are not, therefore, provided as legal advice but as general information for managers. Please consult an attorney for contract development and assistance.)

Offer and Acceptance: an agreement for one party to provide something at a price and an acceptance by the other party of those terms, conditions, and price. That offer and acceptance is binding on both parties.

Identification of the Parties to the Contract: the legal entities who are committing to the agreement. Those parties must have the legal authority to enter into a contract. In public-sector organizations, legal authority to enter into a contract is limited to specific entities identified in statute. Therefore, when a project enters into a contract with a vendor, the contract will usually be put in place between the vendor and a higher-level entity than the agency or the project team. As a result, getting contracts approved in the public sector by those with the authority to bind an agency to a contract can take a considerable amount of time. Public-sector employees and managers should always be aware of the discretion that is allowed to them to sign agreements. In most cases, that discretion will be very limited.

Breach: failure by either party to uphold their part of the deal.

Effective Dates: the beginning and end dates of the contract and, in some cases, the means by which the contract can be extended.

Termination: terms that describe how the contract can be ended. Early termination is termination before the scheduled end date. Termination clauses usually identify the circumstances, such as breach of the contract that can create termination and how either party can terminate the contract. Early termination clauses should identify how costs incurred will be reimbursed.

Conditions: the major terms of the contract.

Representation: the persons with the ability to speak for the parties. Both the buyer and the seller have to know who has the authority to speak for them and the other party with regard to the contract.

Staff persons of public-sector organizations with outsourcing arrangements in place have gotten themselves in trouble by acting as if they were the authorized representative of the agency. Staff should be very careful to understand the mechanisms by which the contract can be changed and who speaks for the organization. They should also make their authority (or lack of authority) clear to vendors in any discussions that deal with the legal documents. The last thing an employee of the buying organization should want is to have the vendor allege that they authorized a change in required vendor performance.

Dispute Resolution: how disputes will be resolved. In some cases, the contract will require arbitration in lieu of litigation. Arbitration is a structured process of dispute resolution in which the dispute is referred to an independent third party rather than the courts. The process for identifying the arbitrator is also usually identified in the contract.

Amendments: the process for making changes in the contract to prevent starting over when changes occur. As noted, creating good contracts that are clear and induce the right behaviors is a particular challenge in an environment where circumstances change often.

Confidentiality: rules for maintaining confidentiality of information. In public-sector contracts, confidentiality can be compromised by open-records laws. Consult an agency attorney to determine requirements and limitations on public information.

Intellectual Property: terms of the contract that identify who owns intellectual property created under the contract. If the vendor creates intellectual property during the course of the contract, the ownership of that property will have to be specified.

For example, the vendor may create software patches specific to the needs of the agency. When the contract terminates, who owns the patches? If they are owned by the vendor, the vendor can resell them to other clients. In some cases, the buyer will not care if the vendor resells their work; in other cases, they will care. Ownership of intellectual property can also influence the amount paid to the vendor. In most cases, intellectual property created under a contract is owned by the customer rather than the vendor.

Fees and Incentives: fees to be paid to the vendor (and the timing of those payments) and the incentives offered. Incentives should be designed to induce vendor behavior that benefits the organization above and beyond the services required. Incentives can be built in for such activities as meeting deadlines early or exceeding customer satisfaction goals. Penalties can also be applied, though the term "reduction in fees" instead of penalties may be better.

Any vendor contract will also contain other legal clauses best left to the legal department. Most large organizations can also provide a template for outsourcing agreements.

The Service-Level Agreement

While the contract sets out the general terms of the agreement, the service-level agreement (SLA) sets out the terms and conditions for the daily operation of the relationship. In those cases where goods rather than services are being purchased, the guiding document is called a statement of work (SOW). The same principles described with regard to SLAs pertain as well to SOWs.

The SLA sets the expectations for both the buyer and provider of services and details the requirements for service delivery at a level that ensures clarity for both parties. A challenge of SLAs is to avoid reversion to describing how work is to be performed. As described in the section on management of outcomes, describing what outcomes will be provided is much better, although elements of what will be provided will also include when and where it is to be delivered.

In the SLA, the project team needs to define what products or services will be provided, timing of delivery, penalties for nonperformance, how disputes are to be resolved, and the full responsibilities of each party, including details of what materials or components are to be provided to the vendor by the project team. Those elements need to be described in detail that can guide and evaluate performance.

A clause in an SLA might state that "the vendor shall provide support to the users of the system 24 hours per day every day of the year." Although that might be our goal, it is a very poor requirement to be included in the SLA. Constant 24/7 support every day of the year is an impossibility. Although we may not be able to identify what will cause services to be disrupted, something will periodically happen. The goal in the SLA is to get the services we need but, at the same time, recognize that problems will occur and create mechanisms for minimizing those problems.

An example of a better-written SLA requirement follows:

> User Support:
>
> The vendor shall provide telephone assistance to agency users of the software via toll-free number. Operators shall be certified user support technicians and shall have completed a 40-hour training session specific to the software. Call waiting times shall not exceed 1 minute. Services shall be provided on a 24/7 basis.
>
> Service outages shall not exceed 2 hours in any one-week period or 2 hours in any calendar month. The vendor shall maintain records of call pick-up times. No more than 5 incidents of slow pick-up are allowed in any calendar week. Total payment to the vendor will be reduced by 10 percent for each month of service if service outages exceed limits.
>
> The vendor shall establish a complaint system and prepare a report detailing the types and frequencies of complaints received. The process for filing complaints shall be online, and users shall be notified of the availability of the complaint system while waiting for call pick-up.
>
> If 10 percent of call-waiting times exceed one minute in any calendar week, total payment to the vendor will be reduced by five percent for each week of service.

Although SLAs vary by application, they should contain several key elements. Those elements include:

- *Definitions of service delivery*: What are the services to be performed and under what conditions should service be available? What metrics are to be used to evaluate service delivery? and What happens if services are not delivered as specified?

- *Support for the service*: When is support to be available and what support is to be provided? What metrics are to be used to determine if support is provided? and What happens if support is not delivered on time and adequately?
- *Special requests*: What happens if a service is requested that is not defined by the SLA? and Are extra services to be paid for on an hourly billing basis?
- *Duties of the buyer*: What support is required to be provided to the vendor by the buyer? and What happens if that support is not provided?
- *Contact persons*: Who is to be contacted in the event of a problem? and How does contact work during normal business hours and after hours?
- *Conflict management*: What happens if conflict occurs? Who is initially engaged? and What documentation and notification are required?
- *Escalation of issues*: How are issues escalated if they are not resolved as required by normal means as specified? What are the second and third steps in escalation? When do legal mechanisms kick in? and Are arbitration or mediation required in lieu of litigation?
- *Dedicated resources*: Does the buyer get to approve significant changes in resources provided to service delivery? and Can the buyer prescribe a reserve margin in staffing?
- *Payment for services*: When will payment be made? How soon after submission of an invoice will payment be made? Who will approve payments? What format do invoices have to be in? and What happens if payment is not made in a timely manner as defined by the agreement?
- *Withheld payments*: What causes payments to the vendor to be withheld? and What mechanisms exist for resolving payment issues? Note that some organizations, based on legal advice, use the term withheld payments rather than penalties.
- *Security*: What security processes are required? and Who has responsibility for the security of records and information?
- *Acts of God*: What outside factors can override the agreement?
- *Time*: What period does the agreement cover? and Can it be renewed?
- *Change management*: How are changes negotiated?

- *Authority*: Who is allowed to speak for the parties and approve minor changes?
- *Knowledge transfer and retention of records*: How is the vendor required to capture knowledge and transfer it to the buyer organization? and Who is responsible for maintaining records and data?
- *Service variance*: Are there other days when services are adjusted?

The project team needs to realize that the SLA represents both a legal document and a means of creating a solid working relationship with the vendor. It is a living document and, at the level of detail included in the SLA, changes will certainly occur.

MANAGING CHANGES AND EXPECTATIONS IN THE VENDOR RELATIONSHIP

Despite the preparations we might make for an outsourced relationship, changes will occur. In fact, one of the reasons that those relationships fail is that neither the buyer nor the seller adequately prepares for changes. We need to remember that the services we have outsourced will be subject to as many changes as the other portions of our business.

No matter how well the documentation for the arrangement with the vendor or partner has been created, that arrangement will be subject to changes. The best agreements anticipate those changes early and build mechanisms for managing changes. Changes can be related to such issues as:

- The number and type of services required
- Organizational processes and tools
- Software and hardware
- Responsible parties on either end
- Contract breaches or failures to provide services
- Costs
- Reporting requirements
- Vendor-provided suggestions for business improvement, which the vendor should be given incentives to provide

The best change management processes are built on a solid foundation of clearly assigned responsibility for managing the vendor relationship.

Although people at various levels of the organization may interface with the outsourcing provider, someone must be in charge of the relationship. Changes must be coordinated and managed by that person or office. As mentioned earlier, having unauthorized persons making changes can create havoc in the management of the relationship.

Change management processes also must be built on a solid set of legal documents that anticipate change and specify change management processes and on a set of clear and realistic metrics for performance. It should be easy to determine if the vendor has met performance requirements. Similarly, unrealistic performance standards offer little guidance for evaluation.

The success of vendor relationships depends on the long-term development of trust in the relationship. That trust will not be in place on the first day of the relationship but can be fostered by frequent communications, including early warning systems that let the team know that problems are brewing.

BEST PRACTICES FOR PUBLIC-SECTOR PROJECT PROCUREMENT MANAGEMENT

Best practices in public-sector project procurement management include:

- Make an ally out of purchasing offices
- Take the time to review purchasing rules and requirements
- Conduct good make-or-buy analyses before relying on vendors for project performance
- Seek the advice of legal counsel for developing the contract with a vendor and for managing that contract
- Make certain that the project team works with the purchasing and legal departments to ensure that project needs are met in purchasing decisions and contract negotiations
- Employ screening criteria for the first screen of vendor proposals
- Employ a bidders' conference to ensure that communications with the vendor are fair and open
- Brief project staff on ethical and legal constraints on their behavior
- Seek to avoid conflicts of interest and the appearance of conflicts of interest
- Focus on outcome management in relationships with vendors rather than activity management

- Make certain that the project manager has a role in approving payments to ensure that project services have been appropriately rendered before payment
- Define services clearly within the SLA
- Do not seek perfection in vendor performance, but realize that problems will occur
- Build processes for dealing with exceptions and problems in the contract and SLA
- Document purchasing decisions and vendor interactions
- Make it clear to the vendor and project team members who has the authority to speak for the project
- Create a solid process for approving and making contract changes

DISCUSSION QUESTIONS

1. What new demands do you face as a manager? What demands are rooted in the expectations of more-senior managers and which are related to your employees or vendors assigned to you?
2. What challenges have you faced in working with your purchasing agency? What strategies have worked?
3. Do you employ a bidders' conference? How do you make certain that all vendors receive the same information?
4. What challenges have you encountered in managing vendor contracts? Have contract disputes reached the point where legal action was required? What was the result and what lessons were learned?
5. What processes have you used to select vendors? Have those decisions been contested by vendors? What role do legal counsel and purchasing department staff play in the purchasing decision?
6. What minimum requirements do you require vendors to meet? What additional minimum requirements do you think might be useful?
7. What types of contracts have you used for your projects? What worked well and what did not? What types of disputes were created?
8. What criteria do you use to determine when to buy goods or services or use in-house resources for your projects? What challenges have you experienced in purchasing goods or services?

9. What activities does your agency engage in? What activities are monitored or measured? Do these activities provide a solid indicator of your effectiveness?
10. What cultural changes do you anticipate in your outsourcing initiatives or use of vendors for major project deliverables? What cultural barriers have you faced and what strategies have you deployed to cope with them?
11. What terms are used to describe the documents that create and record the agreement between your organization and its outsourced vendors?
12. What challenges might your organization face as it makes the conversion from activity to output and outcome management?
13. What are the outcomes of your agency or department? For outsourced operations, what outcomes can be measured?
14. What arrangements or mechanisms do you use for managing changes in your outsourcing relationships? How well have they worked? What improvements could have been made?

EXERCISES

1. Using the template shown earlier in the chapter, create a performance improvement plan for an operation or process. Try to identify measurable output and outcome indicators to measure success.
2. For an outsourcing arrangement you are familiar with, identify the terms you would suggest to your legal department for the outsourcing contract.
3. For a project you are familiar with, identify minimum requirements for vendors and a system for evaluating vendor bids. Identify the criteria that will be used to evaluate bids and the weights to be assigned to each of the factors.
4. For an outsourcing project you are familiar with, identify the components of the SLA that are necessary and create an SLA.
5. For an outsourcing relationship you are familiar with, create a set of mechanisms for managing changes

The Construction and Reconstruction of the Panama Canal

Some exciting and interesting public-sector projects can be "do-overs" of prior failed or even successful projects. One project that fits into both of those categories is the reconstruction of the Panama Canal.

The first attempt to construct a canal connecting the Atlantic and Pacific Oceans through a land connection crossing Central America was begun in 1880 by the French under the direction of Ferdinand de Lesseps. Unfortunately, two problems plagued that project: disease, in the form of malaria and yellow fever, and engineering problems. By the time the project was abandoned in 1893, nearly 22,000 workers had died, and it had become impossible to maintain a skilled project workforce.

President Theodore Roosevelt, however, saw a canal as being critical to U.S. strategic interests. As he attempted to expand the reach of U.S. policy across the world using the U.S. Navy as the enabler of his strategy, he concluded that a canal would be necessary for his plans. When the existing government of Columbia, of which Panama was a part, would not give him the rights to the land necessary for the Canal, a local revolt was organized, and Roosevelt quickly stepped in to support the revolutionaries with military power and funding. The rebels gave the United States unrestricted control of the land necessary for the Canal.

Before the actual construction of the Canal, however, two important deliverables had to be created. The first was an infrastructure necessary for moving huge amounts of soil, which required reconstruction of the Panama Railway. The second was the eradication of mosquitoes from the area, which dramatically reduced the incidence of the two diseases that had halted the original effort. Still, thousands of workers died during the U.S. construction effort.

When it was completed, the Panama Canal was an engineering marvel. The Canal is 48 miles long and passes through three sets of locks that lift ships over the landmass and drops them back down to sea level. By passing through the Canal, ships cut the mileage from New York to San Francisco by over half (6,000 miles instead of 14,000). In 2005, more than 14,000 vessels passed through the Panama Canal.

Impressive as the Canal is, the Canal construction is not over. Ships have continued to grow in size and Canal traffic is increasing. In 2001, projects were undertaken to widen the Canal sufficiently to allow two of the new, wider ships to pass and to straighten some of the curves. In 2002, a project was initiated to deepen the Canal and to increase the water reservoir for the Canal, which will diminish the impact of drought on Canal traffic.

An even more ambitious project is in planning stages. That project will add a third lane to the Canal, which will require the addition of two brand-new lock complexes at each end of the Canal. It is doubtful that the Panama Canal will ever be regarded as complete. It, like other public-sector projects, will continue to be enhanced to meet new, unknown public-interest needs.

Chapter 13

Managing Complexity and Chaos in Public-Sector Projects

Although the project management methods described in prior chapters can help reduce the risk inherent in public-sector projects, they may not be fully adequate for some projects, particularly complex circumstances. In some projects, chaos and complexity work together to frustrate even our best project management methods.

This chapter examines the role of chaos and complexity in public-sector projects and identifies some tools that can help manage uncertainty. In specific, it examines complexity and chaos and what has been learned about it in recent research, looks at the methods for managing chaos and complexity in professional project management, and looks at three tools that might be useful to project managers in public-sector projects as they cope with the turbulence in their projects.

THE ROLE OF COMPLEXITY AND CHAOS IN PUBLIC-SECTOR PROJECTS

Every public-sector project manager knows the feeling. The project has been initiated and planned using the best project management methods, and all of the stakeholders have reached agreement and are committed to the project. The controls are in place, the team is assembled and assignments have been made, and the project is ready to go. Everything that can be done to bring order to the project has been done. It is the calm before the storm, and expectations of success are high.

Most public-sector project managers also know the feeling that comes soon afterward. Barely into project execution, what looked like a sure thing explodes. Rather than being the well-organized effort it was designed to be, the project looks more like a ride straight out of a professional bull-riding event. Instead of managing the project, the project manager finds himself or herself hanging on for dear life with little ability to control outcomes. Instead of watching the project unfold as predicted, chaos ensues.

The truth is that our attempts to bring order to complex projects often come up short. No matter how much planning or organizing we do, we are no match for the turbulence that real-life projects can encounter. Complex public-sector IT projects, military operations, and organizational change projects are particularly subject to devastating disorder.

Complex projects look more like a ride on one of those mean-spirited, massive, and agile bulls than the product of good planning, and though we might try to harness the forces that drive the project and tie ourselves onto the bull as tightly as we can, once that gate opens, all hell can, and usually does, break loose. Like that bull, the project seems to be intent on ruining the day of anyone close to it.

If we are going to be successful on complex projects, we need to see them realistically for what they are and get away from our preoccupation with control. (Recall that we began our transition from an emphasis on control by learning to manage outcomes rather than provide micromanagement of activities.) Fortunately, other disciplines have come to grips with the impact of complexity and chaos, and, if we listen carefully, they can provide lessons for project managers.

That is the purpose of this chapter—to identify what is known about turbulence, chaos, and complexity and identify lessons and methods that can be applied by project managers to mitigate the damage they cause and harness the energy they release. These methods complement the methods and processes required by modern project management standards. In particular, they provide supplemental techniques for communications and risk management. In specific, we will examine three strategies: (1) social network analysis, (2) contextual and relational management, and (3) bifurcation points.

MODERN INSIGHTS INTO CHAOS, COMPLEXITY, AND TURBULENCE

Human beings have always sensed the existence of forces that frustrate their best efforts. In the Native American cosmology, those forces were personified by Coyote, the trickster god. Although variations occur,

Coyote is regarded as powerful, sometimes linked to the creation of the universe. He also can be either mean-spirited or kindly, and is often seen as surviving on his wits and trickery.

The coyote can actually present lessons for the management of organizations. Coyotes are very similar in genetic makeup to wolves, but coyotes can be argued to be a much more successful species. Every attempt to eradicate coyotes has led to their numbers increasing. They now, for example, exist in all 88 Ohio counties, including the most urban. How are they so successful?

They are:

- Extremely adaptable
- Voracious learners
- Unfinished as a species (Coyote DNA is much more variable than wolf DNA.)

In the scientific literature, the same types of forces creating disorder and entropy are recognized, though less colorfully. Complexity theory is concerned with the behavior over time of complex systems. Those systems, according to the various theories of complexity, have a tendency to dissolve into disorder, which in turns exhibits characteristics of a new order. Complexity theory has been applied to such diverse problems as traffic flows, earthquakes, the stock market, the Internet, urban planning, and group dynamics. Chaos theory is a subset of complexity theory and describes the patterns of order that operate under apparently disordered systems.

A theme related to complexity is turbulence. Turbulence is defined as a disordered state of motion within a fluid. It differs from laminar flow, which is smooth and ordered. Turbulent flows are irregular and complex and change constantly. In a laminar flow, there is a strong adhesive force between the flows through a system and the walls of the system. As long as the flow remains under a critical rate of flow, any perturbations in the flow will be quickly passed, and the flow will return to the laminar state. Turbulence occurs when the flow rate exceeds the critical level and the flow becomes unstable, which can happen suddenly. For example, when blood is flowing through arteries in the human body, when it reaches high speeds, it can encounter an obstacle. Turbulence is the result, which in turn results in noise detectable by a stethoscope.

In our projects, turbulence is exacerbated by the combination of the speed at which projects are required to produce deliverables and the many obstacles public-sector project managers need to cope with. At low speeds, we could identify the obstacles as they appear and amend plans to maneuver around them. At high enough speeds, we simply collide with those obstacles and lose our balance. In complex projects, these factors combine to derail or frustrate our efforts to manage project outcomes. Instead of the planned future envisioned in our project documentation, new, unexpected futures develop, and our projects are deemed as failures.

THE CHALLENGES OF CHAOS AND COMPLEXITY FOR PROJECTS AND THE RECOGNITION OF THE LIMITS OF CERTAINTY

Traditional project management has its roots in engineering. Before the current organizational focus on information technology deployment and application development and business change projects, project management was viewed as a discipline that could be employed to organize and track the many details of large-scale projects in engineering and construction.

Traditional project planning has involved a process of creating ever-more extensive arrays of interlocking "if-then" statements and, subsequent to the selection of the right set of statements, controlling the factors that might subvert them. For example:

- If the project is carefully initiated, it can begin with a high level of stakeholder consensus.
- If appropriate scope planning processes are followed, the scope of the project can be determined.
- If people and other resources are available as planned and if team members perform as expected and directed, activities can be performed as planned.
- If the appropriate activities are performed, the predetermined deliverables will be created.
- If the appropriate deliverables are created, project goals will be met.
- If risks are identified and managed and the right risk management strategies identified and planned into the project, the project can progress as planned, although workarounds will be required for unplanned risks.
- If quality standards are identified and built, they can be achieved.
- If stakeholders are identified and the communications plan followed, stakeholders can be satisfied.

To the extent that these if-then statements are correct and produce the intended outcome—and only the intended outcome—the project plan can be stable. To the extent that they are incorrect, incomplete, or produce other outcomes, the project plan will be overwhelmed by chaos. They overlook the fact that there is the potential that project activities will create multiple outcomes, only one of which is the one intended in the project plan.

All project-planning documents are forward-looking and incorporate a predicted vision of the future. Project management, in fact, can be considered an activity that attempts to match good predictions with good organization to create a task organization that will lead to desired outcomes. In a chaotic business environment, however, the ability to predict future outcomes and operate in an environment amenable to deterministic project management practices has been called sharply into question.

Fortunately, there has been substantial recognition within the project management literature and professional practice of project management of the role of uncertainty in information technology projects and other complex projects. Some of the tools of modern project management include the following tools for recognizing and dealing with uncertainty, which will be described in turn:

- Progressive elaboration
- The identification of issues
- Project life cycles
- Stochastic versus deterministic estimating methods
- Ongoing risk identification and analysis
- Change control processes

Overall, modern project management makes a substantial move forward in the recognition of the uncertainty of projects. Although it requires the development of a project plan, it also recognizes the importance of a variety of processes that address the fact that many projects exist in uncertain environments with outcomes that may evolve through the course of the project.

Progressive Elaboration

Project planning is an iterative process that relies on the gradual unfolding of information relevant to the project rather than one-time planning.

This concept is known as progressive elaboration. Progressive elaboration allows the project team to remain open to changes and the emergence of unplanned expectations or outputs. It emphasizes the idea that projects are temporary and unique and undermines the misguided idea that the project manager is "all-knowing" and a completely accurate prognosticator. Progressive elaboration should, of course, be coupled with solid scope definition.

Progressive elaboration allows the project manager and the project team to recognize that plans, while useful, are limited and based on the best information available at the time they are created. It should also alert stakeholders to the fact that the project is an evolving endeavor rather than a static, predetermined set of team assignments.

Issues

Issues are decisions that need to be made or facts in question. They can identify points that project stakeholders disagree about and that will need to be resolved at some point in the project. Issues assert that there are things in the project that are unresolved and that, when they are resolved, they will create project impacts.

Issues, like the concept of progressive elaboration, introduce another point of uncertainty into project management, which complements project risks and assumptions. Issues imply an unfolding project reality that cannot be completely defined at the project's inception.

Project Life Cycles

Project life cycles create a set of standard phases for a project. Performed in sequence, they move the project from its start to its finish. Project life cycles define the work to be done in each project phase and the transitions between phases.

Organizations have moved away from the traditional waterfall-style life cycles that presumed that the needs of project stakeholders could be identified once-and-for-all early in the project and that product developers could then create products in isolation from that point forward. The waterfall life cycles presumed that project functionality and requirements could be identified early in the project and that they would remain relatively constant and stable throughout the project.

Today, many organizations, especially in information technology projects, have adopted replacement life cycles that allow iterative requirements definition, and some have adopted "agile" life cycles. Agile life cycles engage developers and users in a collaborative dialogue designed to iteratively develop user requirements. Those agile life cycles also ensure that changes in user requirements are identified early and factored into product design as smoothly as possible.

The new project life cycles, like the other project management tools being described in this section, provide recognition of the complexity and instability of projects. They give us one more tool for managing changes and the evolution of the project from its start to its finish.

Stochastic versus Deterministic Estimating

Another indication of project management's recognition of unpredictability is the increasing use of stochastic rather than deterministic estimating methods. Deterministic estimates assume we can predict work effort (cost and time) with certainty, and, as a result, deterministic estimates use single-point estimates of activity duration or cost.

Stochastic estimates employ a range of estimates of duration or cost for each activity rather than the single estimate derived in deterministic estimating. Stochastic estimates, as a result, allow us to identify a range of possible outcomes with probability estimates attached to each.

One very simple stochastic estimating technique is the three-point estimating technique. In that estimating technique, an optimistic estimate, a pessimistic estimate, and a midrange estimate are identified for each activity (cost or duration). The total estimate is determined by adding the pessimistic and optimistic to the midrange estimate, which is multiplied by four, and dividing the sum by six. The effect of the multiplier is to buffer the effect of the high and low estimates.

Because stochastic estimates contain a random variable and result in a range of estimates, the project manager can use techniques like Monte Carlo simulations to identify the probability of finishing the project or activity within an acceptable range. Monte Carlo analysis differs from critical-path estimating and management, in that it iterates the project cost or schedule estimates many times to calculate a distribution of possible total project cost or completion dates. The critical-path method, on the other hand, relies on a single, deterministic estimate.

Ongoing Risk Identification and Analysis

One of the most important indicators of the lack of certainty in projects is the concept of risk identification and analysis. Risk is the overt recognition of uncertainty in a project. It looks into the future and identifies the "known unknowns" of the project and seeks to create plans for monitoring those points of uncertainty and responding to them to minimize the overall impact on the project of the negative risks. It seeks to maximize the project impact of opportunities or positive risks.

Good risk management requires the project team to immerse itself in the uncertainty in the project and its environment. It requires constant evaluation of both risks and the assumptions that may have been made very early in the project because of changing circumstances.

Change Management Processes

Change management is driven by the premise that things in projects change and evolve. While the project manager is responsible for preventing unnecessary changes, the project manager must identify necessary changes, explore options, and implement and document those changes. Modern project management builds extensive change management systems and processes into projects. For example:

- Corrective and preventive actions are built into the monitor and control project work-process group.
- Integrated change control is applied from the start of the project to its completion to manage the ripple effect of changes on various aspects of the project plan.
- Configuration management systems are deployed in order to identify and control changes to the product of the project.
- Change control systems provide formal documentation of changes.
- Change control boards bring together stakeholders for formal consideration of change requests.
- Monitoring and controlling processes are established for the scope, schedule, cost, quality, and risk knowledge areas.

In short, modern project management both recognizes the uncertainty inherent in projects and builds processes for identifying and managing the changes caused by the evolution of the project. In order to create successful public-sector projects in a fast-paced, complex environment, however, some additional tools may be required.

FACTORS CREATING COMPLEXITY IN THE PROJECT ENVIRONMENT

We have argued that the deterministic, linear models of project management, which have been at least partly successful for simple projects with clear outcomes, may be inadequate for many of today's projects, which have uncertain outcomes, many diverse stakeholders, and operate in a complex, high-speed project environment. The need for adaptation of project management practices in complex public-sector projects is ably demonstrated by visible failure of some of those projects, failures that cannot be sustained in today's resource-limited, high-demand environment.

Some of the factors that influence the relative complexity of projects and limit the application of traditional management practices are listed in Table 13.1.

Table 13.1 Simple versus Complex Projects

Project Attributes	Simple	Complex
Environmental conditions	Ordered	Chaotic
Stakeholder interactions	Clear and bounded	Emerging and uncertain
Organizational form	Hierarchical, traditional	Network organization
Problem dimensions	Known problem with known outcome at initiation	Wicked problem with uncertain outcome
Level of project manager influence	Able to manage activities and short deliverables, direct control of project personnel	Limited to ability to create broad rules for interactions and performance, indirect
Technology used in the solution	Known	New and unfamiliar
Environmental turbulence	Low levels of turbulence, conflict, and volatility	High levels of turbulence, conflict, and volatility
Project duration	Short, foreseeable future	Longer, stretches into an unknown future

(*continued*)

Table 13.1 (*continued*)

Project Attributes	Simple	Complex
Potential for uncertain events	Low, all variables exist inside the system	High, many outside variables are involved
Organizational impact	Tactical impact only for project outcomes	Strategic impact of project outcomes
Characteristics of systems influenced by the project	Closed, fixed and known	Open, emergent
Organizational rule development	Formal, top-down, articulated in writing	Informal, created by groups, articulated by behavior
Level of knowledge work	Limited involvement of knowledge workers	Full involvement of knowledge workers
Requirements for innovation and group learning	Innovation requirements are limited, solutions and processes are well-defined and tested, required learning is limited and individual	Innovation required for solutions and processes, substantial learning is required and is at the group level as well as individual
Stakeholder interests	Interests known, homogenous, and uniform	Interests unknown, diverse, or in conflict

THREE SUPPLEMENTARY METHODS FOR MANAGING CHAOS AND COMPLEXITY IN PROJECTS

It is one thing to determine that complexity and turbulence are factors that have a high likelihood of impacting project outcomes. It is another entirely to identify and apply methods to manage that complexity and turbulence. Here, we will examine three tools for managing complex projects, tools that can be used to supplement, but not replace, existing project management methods. The tools to be explored are:

- Social network analysis
- Contextual and relational management, including scenario planning
- The identification and management of bifurcation points at which project outcomes are especially sensitive to subtle influences

These tools have a commonality. Each requires a broader systems view of the project and the environment within which it operates. These tools require us to recognize the importance of engaging the project environment in ongoing dialogue and recognizing the nonlinear implications of the intersection between our project and its environment.

Each of the three tools is described in turn in the following sections.

Social Network Analysis

Social network analysis recognizes and maps the relationships among and between people and other network elements. Nodes designate the people or elements and links join them. These relationships operate beneath the formal organization and, often, are unrecognized even though they can substantially impact operations and organizational success. These informal networks are largely unmanaged and evolve naturally. In environments characterized by high speed, complexity, and chaos, they are capable of more rapid adaptation than the formal organizational design.

Although the formal organization designates how actors should interact, social network analysis identifies the critical relationships, how information flows, and how value is created.

A key characteristic of a network is the centrality of nodes. Centrality can be characterized by:

- *Betweenness*: nodes that operate between critical nodes. If a node is a single point of connection between two other nodes, it is important (a broker) but vulnerable.
- *Closeness*: those nodes that have short paths to other nodes. They can connect quickly to other nodes and have significant knowledge of what is happening in the network.
- *Degree*: the number of connections a node has. Although common sense would indicate that the number of connections is the most critical measure of the value of a node to the network, the number of connections is less important than the diversity of the connections. A node with many connections within a tight network of other nodes that are already connected to one another adds little network value.

Other indicators of value in a network are the identification of:

- *Spanners*: those nodes that connect the network to other networks. These nodes can be particularly valuable for gathering information from outside the project or organization and for communicating with external players.

- *Network reach*: the extent to which a network is close or far-flung. Some research has indicated that networks characterized by a multiplicity of close connections (those only one or two steps away) are more functional than dispersed networks.

Social network analysis allows us to identify:

- The most important actors in a network, which can be characterized by their centrality in the network
- How information flows in the network and who controls or expedites that flow
- Critical network nodes, which, if removed or disrupted, can have significant adverse affects on the organization or project
- Isolated network elements that are not well-connected or engaged
- The evolution of the network over time
- The relative centralization (tightness) of the network
- Those nodes that span the network and reach out to other organizations
- Changes in the network in response to crisis

It also requires us to understand the project as a community endeavor.

The Application of Social Network Analysis in Projects

Figure 13.1 illustrates a simple, hypothetical network within a large public-sector information technology project. Although Rick, the project manager assigned from the application development office, represents a network node with many connections, those connections are homogenous and redundant, with the exception of his contact with the PMO director. Rick, though an important player in the project, may not be as critical as Alice.

Alice is connected to fewer nodes than Rick, but Alice represents the only link to a set of nodes, in this case a collection of critical system users. If communications are disrupted with Alice, an entire set of stakeholders will be cut off from project involvement. This breakdown in communications is typical of many information technology projects. Although intense communications continue within the development

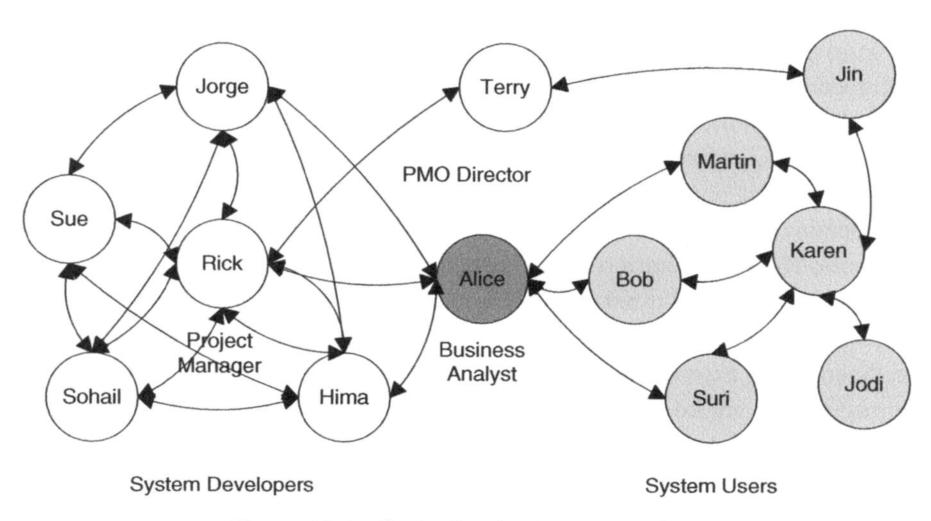

Figure 13.1 Illustrative Project Network

community, communications with the user community are disrupted, and users often become estranged.

In the figure, Karen is a key node relied on by many others in the user community to inform their reactions. If Karen sours on the progress of the project or loses interest in the product being created by the project, her influence will infect others. The project manager and business analyst may want to consider reaching out to her to ensure that her project needs are being met and that she is portraying the project in a positive way.

Network analysis allows us to recognize those links in the project network that are most critical to project success. It also allows us to track the evolution of those networks over time and identify emerging risks and risk triggers. A variety of network analytic tools are available to help project managers identify and track social networks.

Contextual and Relational Management

Contextual and relational management can be viewed as project communications management on steroids. It draws from a variety of sources and theories to focus the project manager's attention on the relationships that exist and evolve between and among project stakeholders and the environment they operate within. It operates on the presumption that projects are social arrangements

These conversations and patterns of relationships occur outside the intentions of project managers and take the form of informal and tacit knowledge exchange that contributes to the establishment and evolution of relationships around the project, relationships that can significantly impact project outcomes.

Because of the evolving social relationships that characterize complex projects, the organization of the project must be engineered for adaptability and change. It must sense changes in its environment and make the requisite changes. As a result, project decision making must be decentralized to the extent possible so that individuals close to stakeholders and changes can make necessary adaptations. Instead of driving project assignments and performance from detailed planning of project activities, necessary project activities must be driven by the evolving needs of the project, needs that cannot be fully defined in advance of the project.

As projects become larger and more complex, the time of the project manager is spent more on managing contexts and relationships than on managing the details and if-then statements of the project. Those relationships requiring the project manager's attention occur inside the project team, but more importantly, outside the project team and at the interface between the project and its external environment.

The goal of contextual and relational management is to engage a wide array of selected stakeholders in an ongoing dialogue about the project, with the understanding that those stakeholders have as much ability to create the project design and outcomes as the assigned project team. It attempts to open the planning and requirements processes and to replace the project team's articulation of a single, future reality as described in the project plan to the co-creation of the future of the project and the outcomes it is intended to create.

Although many complex projects attempt to involve stakeholders by asking for their input into the project's requirements and in order to give them the pretense of participation, techniques for relational management leave open the possibility of a wider set of project futures than the one identified by the project plan, futures that are co-developed by internal and external stakeholders.

The Application of Contextual and Relational Management to a Project

One technique for engaging stakeholders in the creation of a vision of the future is scenario planning. Scenario planning was developed by the planning department at Shell Oil. It recognizes that predicting the future is impossible, but that identification of a set of potential future scenarios could inform project decision making. The results of scenario planning is a set of robust strategies designed to move the organization toward the more desirable scenarios and away from the least desirable ones and to create a set of strategies appropriate for any of the scenarios.

Those scenarios are created in scenario planning by matching pairs of business drivers in a two-by-two matrix to create four scenarios. Those four scenarios are then explored to identify a potential future posed by the scenario. Data can be added to the scenario to quantify the exploration. For example, if we were doing strategic planning for a public-sector agency responsible for providing community health care, we might identify the drivers of our future operations to be:

- The number of clients seeking services
- The availability of funds for treating them
- The ability of the agency to hire competent staff
- The availability of other health care facilities in the community and their willingness to serve our clients

Note that each driver could turn out well or badly. For example, we could see a significant increase or a significant decrease in the number of clients seeking services. At this point in the analysis, the outcome probabilities do not matter.

We can then match any two drivers to create a two-by-two matrix with four potential outcomes. Let's pick the first two on the list and put the number of clients seeking services on the vertical axis, with significantly higher numbers of clients on the top and significantly lower numbers of clients on the bottom. If we put the availability of funds on the horizontal axis, with "adequate funds" on the left and a "shortage of funds" on the right, our matrix would look like the following in Figure 13.2.

In Quadrant 1, we face adequate funding and an increase in clients. In Quadrant 2, we face an increase in clients but inadequate funding. In Quadrant 3, we face inadequate funding, but the number of clients demanding service has declined. In Quadrant 4, we face a decrease in clients but adequate funding.

The technique recognizes that any of these potential scenarios might result, though we might think that some are more probable than others. Our challenge is to:

- Explore each quadrant to identify the impacts on the agency
- Identify strategies for helping steer the environment toward the quadrants that serve clients best
- Identify the strategies that are applicable across the board

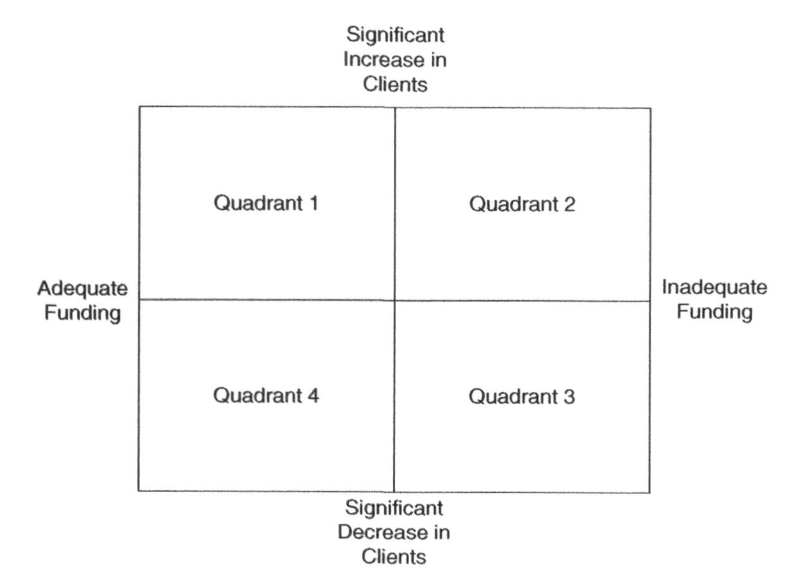

Figure 13.2 Scenario Matrix for Social Services

We can also apply those techniques to project planning. For example, we could engage stakeholders in the identification of scenarios for the implementation of an enterprise-wide information system. The identified drivers of the success of that system might be:

- Stakeholder adoption
- Functionality of the system

Our two success drivers could be matched as shown in Figure 13.3 to identify four scenarios describing potential project outcomes.

Each of these four scenarios could be explored to add detail and to make it tangible. Strategies for each could then be developed, contributing to a final list of accepted strategies. By engaging stakeholders in the exploration of the scenarios, we can build buy-in and create a common vocabulary if we name the scenarios. Scenario planning also helps us to identify potential future problems.

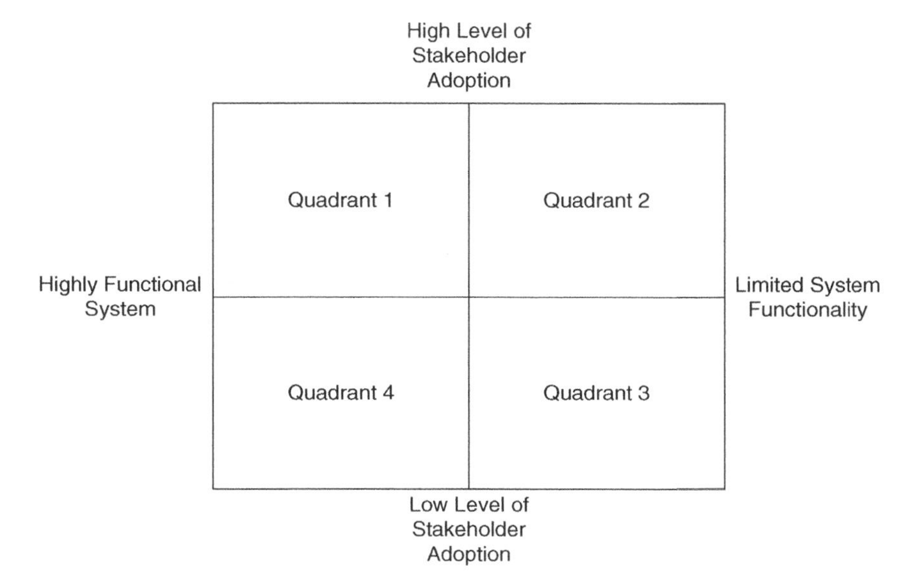

Figure 13.3 Scenario Matrix for Enterprise Information System

Bifurcation Points

A bifurcation point is a point in the project at which a small change in a project variable creates a dramatic change in the projects. Mathematically, a bifurcation occurs when a small, smooth change made to the parameter values of a system causes a sudden change in its long-term behavior. A branch of bifurcation theory is catastrophe theory, which posits that, from a point of bifurcation, a system can either continue to exhibit stability or descend into chaos.

Bifurcation points in projects are like forks in the road. They are the points at which a choice or small change can result in significant changes in the project and even propel the project toward chaos. For example, a complex IT application may require sequential adoption by a set of users. Even though many users may have adopted the system already, we may reach a point at which a key decision leader has to make the choice to adopt or not. If that decision leader elects to reject the system, those users further down the line may also reject it. Some support may erode from those who have already adopted the system. From a decision at that single, unnoticed fork in the road, our system may have descended into chaos.

Similar examples of bifurcation occur in the health care of the elderly. A broken hip in an aged person is a point at which either the individual or his health-care team will be able to isolate the break as a single medical incident and begin the healing process, or that break will begin a cascade of consequences that lead to significant life changes.

The Application of the Management of Bifurcation Points to Project Management

Bifurcation points are those points within a project in which an ordered view of reality, expressed in the project plan, can break down and alternative realities develop. If we can identify project bifurcation points, we can monitor them and take appropriate action to temper their impact. There is a direct relationship between the number of bifurcation points and the complexity of the project.

The first challenge in the management of bifurcation points is their identification. Some might argue that good risk identification and issue identification will simultaneously identify bifurcation points. Although risk and issue identification are not likely to fully identify bifurcation points, the risk register and issue log are good places to start to identify them.

Other likely indicators of bifurcation points would include those points in a project at which:

- Key stakeholders are required to make decisions or commitments.
- Stakeholders interface with the system.
- External stakeholders are in conflict.
- Essential work processes are changing.
- New technology is being introduced.
- There is an impact on highly ritualized processes.
- Current paradigms are being challenged.

Having identified bifurcation points, the project team can give them attention and attempt to influence their outcome.

CONCLUDING COMMENTS ON CHAOS AND COMPLEXITY IN PROJECTS

The project management methods described in this chapter can provide a necessary adjunct to standard project management methods, especially in complex projects. Although standard project management methods can be useful for helping create project dialogue and organizing the work of the project, they provide less-than-adequate analysis and observation of the context and environment within which the project operates.

By maintaining a watchful eye on the emergent patterns in the project environment, the project manager and other stakeholders can adapt the project to the new realities caused by complex and chaotic conditions. By engaging stakeholders in the co-creation of a future state of reality, the project manager can help influence those emergent factors and create the vision of an evolving set of project outcomes.

The methods described in this chapter require skills that are not as prevalent among project managers as might be hoped for. Building a cadre of project managers in an organization who are comfortable with uncertainty and the application of these methods will require more than the standard project management training. It will require training on the application of the new methods, like network analysis, and it will require training programs that change the worldview of most project managers from one that views projects in a deterministic sense to one that regards projects as the creation of unknown and unforeseeable outcomes.

DISCUSSION QUESTIONS

1. What techniques have you used to manage the uncertainty in your projects? What techniques have been embedded in your organization's project management processes?
2. What factors create turbulence in your projects? What analogies can you find between the turbulence created in physical systems when high speed is combined with disruptions and the turbulence in projects?
3. What self-organized systems have you seen develop without guidance or intention? How could those self-organized systems confound your projects? How could you make those self-organized systems work for your projects?
4. What initial conditions might have a dramatic impact later in your projects? How do those initial conditions amplify as you attempt to manage to the project plan?

(*continued*)

5. How would it change your management of a project to view it as a social structure? What techniques would you apply to manage that social structure?
6. What bifurcation points can you identify in your life? What about the point at which your child goes off to college? When your company introduces a new, unproven technology?

EXERCISES

1. For a public-sector project, identify the potential sources of chaos and complexity in it. Identify how you might react to them.
2. For a project you are familiar with, use Table 13.1 to determine its level of complexity.
3. For a project you are familiar with, create a social network map and analyze it. Who are the network nodes that require the most attention? Which nodes represent the most risk?
4. For a project you are familiar with, identify at least two business drivers; choose two of those drivers to create a matrix; name and explore the four scenarios created by that matrix; and identify a set of strategies for the project that are responsive to those scenarios.

Glossary

Acceptance: a risk-response strategy in which it is admitted that the risk may occur but no plan is created to deal with it. Risks are accepted if there is not a good strategy available or if the combination of the probability and the impact of the risk are not particularly high.

Activities (performance management definition): those things people and organizations do to perform their jobs (as opposed to outputs or outcomes).

Activities (project definition): those actions undertaken in order to create project deliverables. Activities are verbs. The list of activities is created as a direct extension of the WBS.

Activity duration estimating: estimating the amount of time (number of work periods) each activity will take.

Activity-on-node (AON): the principle method for creating a project network diagram in which activities are portrayed on the nodes of the diagrams. An activity-on-arrow diagram places the activities on the arrows of the diagram. AON diagramming is also referred to as the precedence diagramming method.

Activity resource estimating: estimating the resources necessary to accomplish the activities identified for the project. A conflict between resources necessary and resources available at the appropriate point in time can require adjustment of the project schedule.

Activity sequencing: placing the identified activities in the proper order in time. Sequencing is driven by dependencies between activities, which can include mandatory dependencies, discretionary dependencies, and external dependencies.

Actual cost (AC): a term used in earned-value management to describe the actual cost of work performed.

Administrative rules: in public-sector projects, mandatory requirements established through structured administrative processes. Rules have the effect of statute and must be created through formal administrative processes that adhere to requirements for notice and participation.

Advertising: publishing the intention of government agencies to contract for goods or services. Advertising increases the available pool of vendors and may be required in many government purchasing systems to ensure that a wide array of vendors is aware of the government's intention.

Analogous estimating: finding a similar activity in the past (an analogy) to see if it provides any clues as to the duration of a current project's activity.

Assumptions: those things regarded as true for the project. Some assumptions make sense and some do not.

Audit: a review of compliance with rules and standards. In project management, a risk audit, procurement audit, or quality audit can be performed to identify how well the project has performed with regard to standards and processes established. In a financial audit, the project records are analyzed to determine the extent to which applicable laws and rules have been complied with.

Avoidance: a risk-response strategy in which steps are taken to completely eliminate the risk.

Baseline: the original project budget, schedule, or scope plus approved changes. The project baselines can change as changes are approved. Earlier baselines can be saved for comparison.

Benefit-cost ratio: the benefits of a project or opportunity and its costs displayed as a ratio. An example is 3:1. Benefit-cost ratios typically are not adjusted to reflect the time-value of money.

Bidders' conference: an opportunity for vendors to meet with government officials to discuss requests for proposals. A bidders' conference ensures that all vendors receive the same information and can be used to limit attempts by vendors to solicit information from government employees.

Bifurcation points: those points in a project when substantial change is possible.

Bottom-up estimating: assigning costs or duration at the lowest level of the project by analyzing the time or cost necessary for individual project activities. Bottom-up estimating requires more data than other estimating methods and is regarded as being more accurate.

Budget: a financial plan for the project. Also refers to the funds allotted to a public agency for its operations over a period of time, usually one year or two.

Budget at completion (BAC): in earned-value management, the original budget for the project.

Business Analysis Body of Knowledge (BABOK®): a compilation of best practices and business analysis methods created by the International Institute of Business Analysts (IIBA®).

Change control: the process of managing the changes to the project. The goal of the project team is to prevent unnecessary changes. However, project changes will occur and must be analyzed, approved, and documented. Those changes must also be reflected in changes to the project plan.

Chart of accounts: the numbering system assigned to deliverables in the WBS and the activities associated with those deliverables.

Claims administration: contract terms that specify how disputes are handled.

Closing: the Process Group concerned with making sure that the project is closed in an orderly manner.

Communications management: the Knowledge Area that ensures that the right information is provided to stakeholders at the right time and in a manner that is useful to them.

Communications plan: a description of who stakeholders are, what they need from the project, and how and when communication with them will occur. The communications plan can be formal or informal, but it should be driven by the needs of the stakeholders, their unique expectations, and the best methods of communicating with them.

Compromise: a technique for resolving conflict that requires a middle ground to be defined so that each party gets something and gives up something. Compromising can be regarded as a lose-lose method of resolving conflict.

Conflict of interest: the opportunity for a government employee to benefit personally from a government transaction. Conflicts of interest are limited by statutes and rules that can contain criminal sanctions on those who violate them. Conflict-of-interest rules can also require government employees to separate themselves from transactions that might create a conflict of interest.

Constraints: the limits on the project, which can include scheduled deadlines, budget, and the available resources.

Context and relational management: examination of the environment of the project and the relationships among stakeholders.

Contract: a mutually agreed-upon arrangement that contains an offer to provide goods or services by the seller and acceptance by the buyer with an agreement to pay.

Contract management plan: documentation of roles and responsibilities and processes for managing project contracts. In public-sector organizations, the contract management plan often involves purchasing administrators.

Corrective action: action taken to address project problems that have already occurred. It is distinct from preventive action. For example, if the project has already fallen behind schedule, corrective action could be taken to adjust future deliverables to correct schedule problems.

Cost accounting: the identification of the costs, both direct and indirect, of an activity or endeavor.

Cost management: the Knowledge Area that ensures that the project is completed within the resources available.

Cost performance index (CPI): in earned-value management, EV divided by AC. A CPI of greater than 1.0 indicates that the project is under budget. A CPI of less than 1.0 indicates that it is over budget as of the date of analysis.

Cost-plus-incentive fee contract (CPIF): a contract type that pays the vendor for actual costs and provides a financial incentive for spending less than a targeted amount and penalizes the vendor for spending over the targeted amount.

Cost-plus-fee contract (CPF): also called a cost-plus-percentage of cost contract (CPPC). These contracts pay the vendor for actual costs plus a percentage of the costs incurred as a fee. CPF or CPPC contracts are unwise in most circumstances and sometimes illegal for use in government because they provide an incentive to the vendor to increase costs.

Cost-plus-fixed-fee contracts (CPFF): a contract that pays the vendor for actual costs plus a fixed fee.

Cost reimbursement contract: a contract that pays the vendor or provider of services for the actual allowable costs incurred, which may include direct and indirect costs.

Cost variance (CV): in earned-value management, EV minus AC. A positive CV indicates that the project is under budget at the point of analysis. A negative CV indicates that it is over budget.

Crashing: adding more resources to critical path activities to compress the duration of the schedule.

Critical path: the longest sequence of activities in the project. A delay of any activity on the critical path will delay the entire project. Activities on the critical path have no slack or float.

Decomposition: the process of breaking deliverables down into subsequently smaller pieces.

Deliverables: the things that the project will produce. They can include the project plan, the ultimate product, and deliverables created along the way. A key to good project management is the complete definition of deliverables before identifying activities. Deliverables are nouns.

Delphi technique: a method of estimating that relies on the iterative judgment of experts who are kept separate from one another.

Dependencies: the relationship between activities. Some activities depend on the completion of other activities. Dependencies can be *mandatory*, which implies that there is no way around them. They can also be *discretionary*, which implies that they should be conducted in a certain order in order to reduce risk.

De-scoping: reducing the scope of a project to save time or money.

Deterministic estimating: estimates of activity duration or cost that apply a single-point estimate for each activity. Deterministic estimating is required for critical path analysis.

Discretionary dependencies: a relationship between activities that allows one to precede the other based on the preferences of the project team. For example, the project team may prefer to complete requirements documentation before beginning a system design. System design could begin before the full documentation of requirements but would run the risk of not meeting requirements.

Duration compression: shortening the project schedule so that it fits the time available. The major tools for compression are reducing scope, adding more resources to key activities, and reevaluating discretionary dependencies. The schedule can be compressed only by working with activities on the critical path.

Earned value: the budgeted cost of the work performed in earned-value management.

Earned-value management: techniques used to link the costs of work performed, the work performed, and the work scheduled to be done. Earned-value management is used to monitor project performance.

Encumbrance: funds set aside for a specific purpose. Amounts to cover future project expenditures can be encumbered to reserve them for future use.

Enterprise environmental factors: all of the internal or external factors that can influence the project, such as organizational culture, regulatory systems, or market conditions.

Estimate at completion (EAC): a new estimate of the total costs of the project derived from analysis of project performance to date. If the variances experienced to date are expected to continue throughout the project, the EAC is calculated using the formula: $EAC = BAC/CPI$. If variances are not expected to continue, the EAC is calculated using the formula: $EAC = AC + (BAC - EV)$. If the original budget estimate is viewed as having been proven to be inadequate, a completely new EAC can be calculated using the formula: $EAC = AC + ETC$ (where the ETC is a new estimate of the cost of the work yet to be performed).

Estimate to completion (ETC): an estimate of the cost of completing the remaining work of the project. ETC is added to AC to create EAC if the project team concludes that the original budget is no longer valid.

Executing: the Process Group concerned with performing the work of the project in conformance with the project plan.

Executive order: a directive issued by an executive under authority provided by legislation or a constitution in order to implement a policy. Executive orders have the force of law but must proceed from a clearly identified executive authority.

Expectancy theory: the theory, advanced by Victor Vroom, that suggests employees will perform if they can expect to receive adequate rewards.

Explicit knowledge: knowledge that can be documented and transferred in writing. Explicit knowledge is contrasted to tacit knowledge.

Fast tracking: revising the schedule to perform sequential activities at the same time. Activities that are candidates for fast tracking have discretionary dependencies rather than mandatory dependencies.

Fixed-price contract or firm-fixed-price contract (FFP): a contract that pays a preset price for goods or services. The seller or provider of the service bears the price risk of the transaction.

Fixed-price-incentive-fee contract (FPIF): a fixed-price contract in which the vendor can earn additional compensation for meeting predetermined performance targets, such as completion dates. Highway construction contracts often involve incentives for early completion.

Float: see *slack*

Force: a technique for resolving conflict that requires the project manager to exert his or her authority and to force a solution. Forcing is not regarded as successful in the long term. It does not draw on the full talents and diversity of the team and can result in hurt feelings and animosity.

Freedom of Information Act (FOIA): a U.S. statute that allows citizens to request and be provided information held by government.

Fund accounting: a system of accounting required to be used by government entities to segregate funds into accounts for the purposes for which they were appropriated. Funds can be restricted for specific uses. An exception is the general fund, which is used for general government purposes.

Human resource management: the Knowledge Area that ensures that the project's human resources are used optimally.

Hygiene factors: according to Frederick Herzberg, job attributes that do not contribute to employee motivation, including pay, working conditions, and relationships with peers.

Initiating: the Process Group concerned with starting the project in an orderly manner.

Integration management: the project Knowledge Area that ensures that all of the project work is coordinated.

Interests: what stakeholders actually need as opposed to what they say they need. Interests represent the root causes of the conflict rather than the symptoms of it.

Internal rate of return (IRR): an indicator of relative project value, which incorporates the time-value of money. The IRR is the discount rate applied to a project or option such that the net-present value of the project is zero. A project with a higher IRR is preferred to a lower IRR. A threshold rate or hurdle rate for IRR can also be set for projects; projects with an IRR lower than the threshold rate are not accepted.

IPECC model: the five project management process groups that include project Initiation, Planning, Execution, monitoring and Control, and Closing.

Ishikawa diagrams (also called fishbone diagrams or root-cause diagrams): a technique for identifying the root causes of problems. Ishikawa diagrams resemble a skeletal fish and attempt to identify the actual causes of problems. The logic used in the creation of Ishikawa diagrams is similar to the logic of repeatedly asking why a problem has occurred.

Issue: something in dispute or under question in a project. Issues can include future decisions that need to be made and may involve different opinions. For example, an issue in a project could be whether to build the software in-house or to use off-the-shelf software.

Issue log: a document used to identify issues, their intended resolution, the person responsible, due dates, and priorities. An issue log must be continually updated.

Knowledge Areas: the nine areas of project management that the project manager and team must master (integration, scope, time, cost, quality, human resources, communications, risk, and procurement).

Knowledge management: the deliberate attempt to capture and reuse knowledge in an organization. Knowledge management is particularly critical for public-sector organizations as the members of the baby boom retire and take critical knowledge with them.

Lag: a delay required between successor and predecessor activities that normally would be conducted in direct sequence (i.e., when A is done, B begins). A simple example is the case in which, even though the foundation of the house has been poured, building the frame on that foundation must wait for the concrete to set.

Lead: the opposite of a lag. A lead implies that a successor task can begin before its predecessor is completely finished. For example, basic system design can begin when some of the requirements have been developed and while the remaining requirements are identified and analyzed.

Lean management (lean government): the application of structured processes for improving processes and reducing wasted time or resources. Lean government has proven successful in reducing cycle times and costs for such processes as issuance of licenses or permits. Lean management utilizes some of the processes of Six Sigma but requires less statistical analysis.

Make-or-buy analysis: the determination of whether goods or services can be created or performed with in-house resources versus purchased from vendors. Make-or-buy analysis is used in preparing the procurement plan for a project.

Mandatory dependencies: relationships between activities that require that one precede the other. For example, a system has to be built before it can be deployed.

Mathematical methods: techniques for evaluating projects that incorporate sophisticated mathematical models (e.g., linear programming).

Milestone: a significant event in the project schedule. Milestones can also be thought of as activities in the schedule without any duration. Milestones can provide opportunities for schedule review and determination of schedule adherence.

Mitigation: a risk-response strategy in which steps are taken to reduce either the probability of the risk occurring or the impact on the project if it does, or both.

Monitoring and controling: the Process Group concerned with ensuring that the work of the project conforms to the project plan.

Motivating factors: according to Frederick Herzberg, those factors that actually contribute to employee motivation, including the opportunity for self-actualization and growth. Motivating factors are distinguished from hygiene factors, which include pay, the work environment, and relationships with peers.

Net present value (NPV): the benefits of a project minus its costs, where both benefits and costs realized or incurred over a period of time are discounted to the present for accurate comparison. Future streams of either costs or benefits are discounted using an appropriate risk-based discount factor for comparison at a common point in time, usually the present. A positive indicates that the benefits of a project outweigh its costs. Net present value analysis is superior to payback or benefit-cost ratios for project evaluation and simpler than IRR.

Network analysis (communications): the analysis of the relationships among project stakeholders in order to identify critical communication paths and needs.

Network analysis (project scheduling): analysis of project network diagrams to identify the critical path and other attributes of the schedule.

Organizational process assets: those things available to assist with projects, such as the library, computing assets, project management methods and books, and advice from instructors.

Outcomes: products or services of the organization that are of direct consequence to stakeholders (as opposed to activities or outputs).

Outputs: what the organization produces (as opposed to activities or outcomes).

Payback: a means of evaluating projects by determining the number of time periods required to recoup the investment in the project. Payback typically does not take into account the time-value of money (i.e., future flows of funds are not discounted to the present).

Performance management: identifying the current state of affairs and targeted state of affairs in the way that an organization or individual performs and building a plan to close that gap.

Phase: a set of project activities that result in creating a project deliverable. Phases simply help organize the project. At the end of a phase, the deliverable can be reviewed and accepted or sent back for more work.

Planned value (PV): in earned-value management, the budgeted cost of work scheduled. PV is the summed cost of the cost budgeted for work scheduled to be performed at a point in time.

Planning: the Process Group concerned with creating a solid project plan.

Portfolio: a collection of unrelated projects managed in a coordinated manner. For example, at the organizational level, the project portfolio may contain information technology projects, organizational development projects, process improvement projects, and others.

Positions: the demands or assertions of people involved in conflict. Positions may not accurately reflect the real interests of the stakeholders. The job of the project manager is to identify interests behind the positions.

Position authority: the ability to influence team members as a result of your formal designation as project manager. Position authority is rare in student projects.

Predecessors: an activity that needs to be completed before the next activity (i.e., its successor) because of mandatory or discretionary dependencies.

Preventive action: action taken to address problems that have not yet occurred. For example, if it appears that the project may fail to meet a target, preventive action could be taken to transfer additional resources to the project to increase the likelihood of meeting the target date.

Principled negotiations model: the model of conflict management created by Roger Fisher and William Ury that focuses on separating people from problems, identifying interests, seeking win-win solutions, and using objective criteria. The principled negotiations model is an improvement on traditional positional negotiations.

Process: as distinct from a project, a repetitive activity that creates similar results each time it is initiated. Processes in public-sector projects can include issuing of licenses, tax collections, and recording financial transactions. Improving a process requires a project.

Procurement management: the project Knowledge Area that ensures that goods are purchased, if necessary, from outside the project organization.

Procurement management plan: the output of procurement planning and a plan that describes what will be purchased and how.

Progressive elaboration: the fact that more is learned about a project as it progresses. In early stages of a project, participants have to work with the information available with the understanding that things will get clearer later.

Program: a collection of related projects managed in a coordinated manner. For example, the information technology department may create a program for upgrading systems. That program may contain several individual projects.

Project: an initiative with a start and a finish designed to create something unique. Projects differ from processes, which are designed to create repeatable outcomes.

Project charter: the document used to capture an understanding of the project at its outset.

Project initiation: the process of getting the project off to an orderly start. In project initiation, participants need to make sure they have a good understanding of the project and that others share that understanding.

Project Management Body of Knowledge (PMBOK® Guide), A Guide to the summary of project management processes and methods compiled by the Project Management Institute (PMI). The current version is the *Fourth Edition*.

Project management team: a subset of the project team that has management responsibility for the project. *The leader of the project management team is the project manager.*

Project manager: the person responsible for attaining project objectives.

Project network diagram: a means of illustrating the sequence of project activities.

Project phase: see *phase*

Project scope: what is in the project and what is not in the project, which is defined by the list of deliverables that the project is intended to produce. There can be both product-related deliverables that are handed off to clients and project-based deliverables that help manage the project.

Project team: the individuals with responsibility for creating project deliverables under the direction of the project manager.

Public sector: the portion of the economy composed of government organizations that exist to serve the public interest. Not-for-profit organizations formed to service public interests but that are not created by statute are sometimes included within the definition of the public sector but often are regarded as part of the "third sector."

Qualified sellers' list: a list created by the organization of those vendors who have been preregistered and are eligible to bid on proposals.

Qualitative risk analysis: the identification of risk priorities using probability and impact assessment.

Quality assurance: the general activities taken to ensure that the project meets quality standards.

Quality control: the activities undertaken to ensure that specific project deliverables meet quality requirements.

Quality management: the project Knowledge Area that ensures that the project meets quality standards established for it.

Quality management plan: a portion of the project management plan that describes the quality standards for the project, how those standards will be met, and other related aspects of the project's approach to quality. The quality management plan can be informal or formal and may employ standard quality approaches adopted by the organization.

Quantitative risk analysis: the application of statistical and mathematical techniques to evaluate overall project risk and the probability of achieving schedule and budget targets. Monte Carlo analysis is a tool of quantitative risk analysis.

Random errors: deviations in process results that are within the boundaries of process control and not caused by special causes. Random errors are part of normal process variations. Random process errors are to be expected and not the target of process controls. Random errors are sometimes referred to as common cause errors.

Redundant communications: the practice of assigning two responsibilities to the person sending a message (sending it clearly and making sure it was understood) and two responsibilities to the person who received it (listening carefully and confirming that he or she understood the message correctly).

Request for proposal (RFP): a document soliciting vendors for bidding on work to be contracted. Sometimes called an invitation for bid (IFB) or a request for quotation (RFQ).

Requirements: those deliverables of a project that describe the attributes of the product or service to be provided to stakeholders.

Requirements documentation and analysis: the process of further analyzing requirements through the use of diagramming and modeling techniques. Requirements analysis and documentation allow the project team to ensure that requirements have been fully identified and create a requirements package that can be used to communicate requirements to the group or person creating deliverables.

Requirements elicitation: the process of acquiring the requirements of the project, which usually refers to the requirements for project deliverables to be provided to customers. Requirements elicitation can employ such methods as job shadowing, interviews, document analysis, and surveys. According to the *BABOK*®, requirements elicitation results in the creation of the requirements list, which includes user and supplementary requirements.

Requirements management plan: a plan for how the project team will elicit, document, analyze, communicate, and validate the project's requirements.

Risk: an uncertain event that, if it happens, can impact a project. Risks can have a positive impact on the project, but risk analysis is usually concerned with those events or possibilities that might negatively impact the project.

Risk categories: a general category of risks that can help structure risk brainstorming.

Risk factor (RF): the product of the probability and impact scores for a risk. Risk factors can be displayed using a probability and impact matrix.

Risk management: the Knowledge Area that ensures that project risks are identified, analyzed, and dealt with.

Risk register: a report used to document what is known about the project's risk. Detail is added to the risk register as it is determined. At first it just includes risk categories. Then the risks that have been brainstormed are added. Then the risk probability and impact are added.

Risk scale: a ranking of risks based on the product of the probability of the risk and a ranking of its impact.

Risk tolerance: an individual's relative preference for risks. Some people have high risk tolerance. Others have low risk tolerance. Risk tolerance can vary for types of risk.

Risk triggers: those events that indicate that a risk has occurred or that it is about to occur.

Rules: see *administrative rules*

Secondary risks: risks created by risk-response strategies. Secondary risks do not exist until another risk-response strategy is adopted.

Schedule variance (SV): in earned-value management, EV minus PV. A positive SV indicates that the project is ahead of schedule. A negative SV indicates that it is behind schedule at the point of analysis.

Schedule performance index (SPI): in earned-value management, EV/PV. An SPI of greater than 1.0 indicates that the project is ahead of schedule. An SPI of less than 1.0 indicates that it is behind schedule.

Scope: see *project scope*

Scope creep: the constant addition of new deliverables to a project.

Scope definition: the process of creating a clear project scope.

Scope management: the Knowledge Area that ensures that the project performs *all of* the work that is necessary and *only* the work that is necessary.

Screening criteria: the minimum requirements of vendors bidding on a proposal. Screening criteria can include items such as licenses to perform business, the number of years in business, and insurance or indemnity bonds.

Service-level agreement (SLA): a document describing the services to be performed by a vendor. An SLA is opposed to a statement of work (SOW), which identifies goods to be provided by the vendor.

Slack (also referred to as float): the amount of time that an activity can be delayed without delaying the entire project. Activities on the critical path have zero slack.

Smoothing: attempting to resolve conflict by calming the situation. Smoothing does not address the roots of the problem and often just postpones the conflict.

Special cause variations: errors in processes that can be attributed to identifiable problems. Special cause variations are of concern to those attempting to improve processes, because special cause variations can be acted on and reduced. Random variations or errors cannot be addressed.

Stakeholders: those people or groups who are involved in a project, who are affected by it, or who can influence the outcome. Stakeholders need information about the project. It is the project manager's responsibility to identify what those needs are and provide it in a way that is useful to the stakeholders.

Statement of work (SOW): a list of things to be provided by a contractor (as distinct from the service-level agreement).

Statute: a law passed by an authorized legislative authority that is binding on the project. Statutes usually require the development of rules for their implementation. Both must be adhered to by the project team, and the project team must be aware of all relevant statutes and rules.

Stochastic estimating: creating a range of cost or time estimates for each activity.

Successors: activities that follow others (i.e., their predecessors) due to mandatory or discretionary dependencies.

Theory X employees: according to Douglas McGregor, those employees who must be constantly supervised and controlled in order to ensure performance.

Theory Y employees: according to Douglas McGregor, those employees who are self-motivated and who require coaching and mentoring rather than hands-on management.

Time and materials contracts: a contract that pays the vendor an hourly rate for services provided plus reasonable incurred costs.

Time management: the Knowledge Area that ensures that the project is completed on time.

Total quality management: a strategy that attempts to embed quality into the processes of an organization through continuous, focused attention on the achievement of quality objectives.

Transference: a risk-response strategy in which the risk is shifted to someone else, usually through some sort of financial arrangement.

Transforming: a stage of team development that occurs after the work of the project is completed and the team moves on to other projects.

Triple-constraint model: the idea that the project manager must make trade-offs among project cost, schedule, and quality or scope. A change in one often creates a change in another.

Three-point estimating: adding the pessimistic estimate, the optimistic estimate, and four times the middle estimate together and dividing the result by six to derive a reasonable time estimate.

Weighting criteria: scales used by evaluators of proposals to assign scores to vendor proposals. Weighting criteria can be used by several different evaluators to reduce some of the subjectivity of the evaluation of vendor bids.

Workaround: a reaction to a risk that was not anticipated or planned for.

Work Breakdown Structure (WBS): a hierarchical description of project deliverables. Deliverables are nouns.

Index

Printed and bound by CPI Group (UK) Ltd, Croydon, CR0 4YY

23/04/2025

14660925-0005